Revolution in Measurement:

Western European Weights and Measures Since the Age of Science

Revolution in Measurement:
Western European Weights and Measures Since the Age of Science

Ronald Edward Zupko

The American Philosophical Society
Independence Square Philadelphia
1990

Dust jacket illustration: Elizabeth I, Exchequer standards of the bushel, gallon, quart, and pints, 1601–1602. Used with the permission of the Science Museum, London.

Library of Congress Catalog Card No: 89-84666
International Standard Book No: 0-87169-186-8
International Standard Serial No: 0065-9738

To
My Students
Past and Present

Contents

Contents

Figures

Acknowledgments

Many individuals and institutions made important contributions during the research and writing phases of this book.

First, I am deeply grateful to the Institute for Advanced Study, Princeton, New Jersey, for granting me membership in the School of Historical Studies, and especially to Professor Marshall Clagett for his unstinting encouragement and intellectual stimulation. My sojourn at the Institute enabled me to initiate the concepts upon which this book is based.

In England, Edward Telesford of the Photographic Service, The British Museum, provided microfilms and photographic reproductions of hundreds of European metrological documents and monographs. H. Barrell and Julia B. Johnson of the National Physical Laboratory, Teddington, Middlesex, supplied me with pertinent information on, and several photographs of, current British and metric physical standards. Dr. Barrell, a former Superintendent of the Standards Division, and Ms. Johnson, of the Publicity Section, also helped me locate the sources for additional photographs, which I eventually received from John Moss, Windsor, Berkshire, a member of the American Society of Magazine Photographers, and from the Science Museum in London. A. F. Constantine, the Divisional Chief Technical Officer of the British Standards Institution, gave me several references to technical institutes that were able to supply additional data on the European pre-metric and metric systems. The staff of the Public Record Office in London also performed numerous services.

In Canada and the United States, Marion E. Brown, Head of the Department of Rare Books and Special Collections, The University of Toronto Library, and William H. Patch, Head of the Circulation Department, and Felix Pollak, Head of the Department of Rare Books, both of the University of Wisconsin Memorial Library in Madison, assisted me in locating many valuable early modern scientific and technological texts. The Microfilm Department of the University of Wisconsin–Milwaukee Library helped me on many occasions during the months devoted to working among the thousands of books in the Short Title Catalogue Collections. Three members of the National Bureau of Standards in Washington, D.C., A. G. McNish, L. J. Chisholm, and Ross L. Koeser, extended their resources and expertise on several occasions.

At The National Museum of American History (formerly the Museum of History and Technology), Smithsonian Institution, recognition must be extended to Jon Eklund, Curator of Chemistry and Metrology, for his continuous encouragement and helpful suggestions; Frank A. Pietropaoli, Librarian, for numerous professional services; and the staff of the Dibner Library for allowing me to study hundreds of rare scientific and technological works within their peaceful premises.

I am also in debt to various personnel at the following institutions for their assistance: Bibliothèque Nationale, Paris; Library of the Pontifical Institute, Toronto; Library of Congress, Washington, D.C.; The New York Public Library, New York City; The University of Chicago Library, The John Crerar Library, and The Newberry Library, Chicago; Princeton University Library, Princeton; and Marquette University Memorial Library, Milwaukee.

Finally, I wish to thank the National Science Foundation, the American Philosophical Society, and the Committee on Research of Marquette University for many fellowships and research grants, and Dennis Mueller and Patricia Windels, my research assistants, for working patiently, faithfully, and diligently. At Marquette University, Dr. Lynn E. Miner, Assistant Dean of the Graduate School and Director of Research Support, Dr. Thomas E. Hachey, Chairman of the Department of History, and Rev. John P. Talmage, Head Reference Librarian, assisted me in many different and valuable ways over the last decade.

Milwaukee, Wisconsin
1989

Ronald Edward Zupko

CURRENT METRIC WEIGHTS AND MEASURES

Linear Measures

millimeter	0.03937 in
centimeter	0.3937 in 0.03281 ft 0.01094 yd
decimeter	3.937 in 0.3281 ft 0.10936 yd
meter	39.37 in 3.2808 ft 1.0936 yd
dekameter	393.7 in 32.808 ft 10.9361 yd
hectometer	328.084 ft 109.361 yd 19.8838 rd
kilometer	3280.8 ft 1093.6 yd 0.62137 mi

Area Measures

square millimeter	0.00155 sq in
square centimeter	0.155 sq in 0.00108 sq ft
square decimeter	15.5 sq in 0.10764 sq ft 0.01196 sq yd
square meter	1550.0 sq in 10.7639 sq ft 1.196 sq yd 0.0395 sq rd
are	1076.391 sq ft 119.599 sq yd 0.02471 acre
hectare	107,639.1 sq ft 11,959.9 sq yd 395.367 sq rd 2.471 acres
square kilometer	247.105 acres 0.3861 sq mi

Capacity or Volume Measures

milliliter	0.061025 cu in 0.007 UK gi 0.0084535 US gi
centiliter	0.61025 cu in 0.07 UK gi 0.084535 US gi 0.0176 UK pt 0.021134 US liq pt 0.018162 US dry pt
deciliter	6.1025 cu in 0.7 UK gi 0.84535 US gi 0.176 UK pt 0.21134 US liq pt 0.18162 US dry pt
liter	61.025 cu in 1.7598 UK pt 2.113376 US liq pt 1.816166 US dry pt 0.8799 UK qt 1.056688 US liq qt 0.908083 US dry qt 0.21998 UK gal 0.26418 US gal 0.0275 UK bu 0.02838 US bu
dekaliter	8.799 UK qt 10.56688 US liq qt 9.08083 US dry qt 2.2 UK gal 2.64172 US gal 0.27497 UK bu 0.28378 US bu
hectoliter	22.0 UK gal 26.4172 US gal 2.7497 UK bu 2.83776 US bu
kiloliter	220.0 UK gal 264.172 US gal 27.5 UK bu 28.37759 US bu

Current Metric Weights and Measures

Weights	
milligram	0.01543236 gn
centigram	0.1543236 gn
decigram	1.54324 gn
gram	15.43236 gn
	0.77162 scr
	0.64301 dwt
	0.56438 dr av
	0.25721 dr t
	0.035274 oz av
	0.032151 oz t
	0.0022 lb av
	0.00268 lb t
dekagram	154.324 gn
	5.64383 dr av
	0.35274 oz av
	0.32151 oz t
hectogram	1543.24 gn
	3.527 oz av
	3.215 oz t
kilogram	15,432.36 gn
	2.204623 lb av
	2.679229 lb t
metric ton	2204.623 lb av
	1.1023 short ton
	0.9842 long ton

Revolution in Measurement:

Western European Weights and Measures Since the Age of Science

1
LEGACY OF THE PAST: THE WEIGHTS AND MEASURES OF PRE-METRIC EUROPE

Most of the modern world's commerce, industry, technology, manufacturing, agriculture, and other business and professional pursuits depend upon the scientifically structured metric system of weights and measures. Consisting of very few legal units, remarkably accurate physical standards, absolutely precise definitions and interunit coherence, and marked by decimal simplicity, metric metrology represents the finest achievement in man's endless quest to measure and weigh all of the objects found in his daily existence. Its universal adoption and application are of inestimable value to our modern way of life.[1]

These same accolades certainly cannot be ascribed to medieval and early modern European metrologies. Characterized generally by confusion and complexity, and dominated largely by custom and tradition, pre-metric weights and measures evolved usually on local or regional bases and were geared to needs profoundly different from those of the modern world, or even those of ancient Roman civilization where Roman standards served as a unifying link throughout the Mediterranean world. In the Early Middle Ages native Celtic metrologies, long laying dormant in the countrysides or hinterlands during centuries of Roman occupation, together with those introduced by hundreds of conquering Germanic tribes, slowly supplanted the weights and measures of Rome. After the turn of the millennium metrological growth and proliferation set in and gathered rapid momentum during the Later Middle Ages due to economic development, commercial competition, demographic growth, increased urbanism, taxation manipulations, transportation refinements, technological progress, territorial expansion, and the continuous impact of custom and tradition. Tens of thousands of new units were introduced and hundreds of thousands of local variations emerged from the Atlantic seaboard to central and eastern Europe.[2] There were more than a dozen principal methods by which unit variations arose prior to the creation and dissemination of the metric system in the late eighteenth and early nineteenth centuries. It should be noted that there were other causes for metrological proliferation as well, but the items presented here were responsible for the greatest number and range of new units over time, especially in the British Isles, France, Italy, and the German states.[3]

Central Governments

Central governments contributed substantially to weights and measures proliferation by promulgating several national standards for individual units that had widespread usage throughout their respective domains. In France, for example, the arpent was the principal measure of area for land, but there were three official standards. The "arpent de Paris" contained 100 square perches, each perche of 18 pieds in length. It was a square whose four sides were 180 linear pieds each, totaling 32,400 square pieds (34.189 a). The "arpent des eaux et forêts" contained 100 square perches, each perche of 22 pieds in length; its four sides were 200 linear pieds each, totaling 48,400 square pieds (51.072 a). The "arpent de commun"—authorized for use in the provinces—was 100 square perches, each perche of 20 pieds. This square had sides of 200 linear pieds each, totaling 40,000 square pieds (42.208 a). The corde, a measure of volume for firewood, also had three national standards: the "corde des eaux et forêts" was a pile 8 pieds long, 4 pieds high, each billet being 3 pieds, 6 pouces in length, or 112 cubic pieds (3.839 cu m) in all; the "corde de port" was a pile 8 pieds long, 5 pieds high, each billet being 3 pieds, 6 pouces in length, or 140 cubic pieds (4.799 cu m) in all; and the "corde de grand bois" was a pile 8 pieds long, 4 pieds high, each billet being 4 pieds in length, or 128 cubic pieds (4.387 cu m) in all. In the late eighth century the perche was fixed under Charlemagne at 6 aunes or 24 Roman pieds (ca. 7.09 m) and this remained the national standard until the end of the Middle Ages when it was replaced by three other perches: the "perche de Paris" of 3 toises or 18 pieds (5.847 m); the "perche de l´arpent commun" of 20 pieds (6.497 m); and the "perche des eaux et forêts" of 3 2/3 toises or 22 pieds (7.146 m).

In the British Isles the situation was just as confusing and complex. A Scots acre, for instance, consisted of 4 roods (ca. 0.51 ha) or 6150.4 square yards or larger than the English statute acre of 4840 square yards (0.405 ha) by slightly more than 25 per cent. The Scots standard pint for liquids and dry products contained 103.404 cubic inches (ca. 1.70 l) or 2 choppins or 4 mutchkins, but the English pint for dry products contained 33.6 cubic inches (0.551 l) and the pint for liquids contained 4 gills or 28.875 cubic inches (0.473 l) for wine and 35.25 cubic inches (0.578 l) for ale and beer. The Scots cloth ell was 37 inches (ca. 0.95 m) equal to approximately 37 1/5 English inches. A Scots furlong consisted of 40 falls (226.771 m) or 240 ells equal to 744 English feet. In Scotland the gallon was employed chiefly for wine and contained 827.232 cubic inches (ca. 13.60 l). This was equal to 3.5811 English wine gallons. The Scots mile was 320 falls (1814.170 m) or 1920 ells equal to 1984 English yards or 5952 English feet. The tun of wine was standardized in Scotland at 60 gallons (ca. 8.16 hl); the English wine tun contained generally 252 gallons (ca. 9.54 hl). Most of the capacity measures for liquids and dry products in Ireland were smaller than their English prototypes. In Wales the variances were not as significant.[4]

Scotland was also unique in establishing its medieval standards for many capacity measures on the weight content of river water poured into certain vessels. The boll, first standardized under David I at 12 gallons or the capacity of a vessel 9 inches deep and 72 inches in circumference, was commonly regarded throughout the Middle Ages as any vessel capable of holding 164 pounds of the clear water of Tay. By 1600 it was fixed at 4 firlots or 8789.34 cubic inches (1.441 hl) and equal to 4.087 Winchester bushels for wheat, peas, beans, rye, and white salt, and 12,822.096 cubic inches (2.101 hl) and equal to 5.963 Winchester bushels for oats, barley, and malt. Both bolls were equal to 16 pecks or 64 lippies. The firlot of Edinburgh was the standard after 1600 for wheat, peas, beans, rye, and white salt, 21 1/4 pints (3.612 dkl) or 103.404 cubic inches each or 2197.335 cubic inches in all and equal to 1.0218 Winchester bushels, while the Linlithgow firlot was the standard after 1600 for barley, oats, and malt, 31 pints (5.270 dkl) or 3205.524 cubic inches and equal to 1.4906 Winchester bushels. Prior to 1600 the firlot was defined as a vessel holding 41 pounds of the clear water of Tay. The Scots gallon for liquids and dry products contained 827.232 cubic inches (ca. 13.60 l) or 4 quarts, 8 pints, 16 choppins, 32 mutchkins, or 128 gills. Throughout the Middle Ages it was defined as a vessel capable of holding 20 pounds and 8 ounces of the clear water of Tay. Finally, the pint of post-1600 vintage of 103.404 cubic inches (ca. 1.70 l) was defined in medieval Scottish legislation and in several acts thereafter either as 2 pounds and 9 ounces of the clear water of Tay, or as 2 pounds and 9 ounces troy weight of clear water, or as 3 pounds and 7 ounces troy weight of water from the river of Leith. The daily or yearly water purity of the two rivers must have caused medieval Scotsmen untold problems.

Regionally, the problem of multiple standards was, of course, even more acute, and two examples can suffice as illustrations. The canne was the principal measure of length in southern France for textiles, but the canne of Marseille was 8 pans or 64 menus or 72 pouces or 892.22 Parisian lignes (2.013 m), the canne of Montpellier was 8 pans or 881 Parisian lignes (1.987 m), and the canne of Toulouse was 8 pans or 64 pouces or 796.2 Parisian lignes (1.796 m). In Paris the "quintal poids de marc" or hundredweight was 100 livres (48.951 kg), but regionally the "quintal toulousain" was 104 livres (42.422 kg), the "quintal lyonnais" was 100 livres (41.876 kg), and the "quintal poids de table" at Marseille and the "quintal d´eau-de-vie" at Montpellier were each 100 livres (40.79 kg).

These standards also changed in size over time. In France, for example, the livre was the principal unit of weight. During the late eighth century the "livre esterlin" was fixed at 5760 grains (367.1 g) and consisted of 20 sous or 12 onces or 240 deniers or 480 oboles. In the middle of the fourteenth century the government of King John the Good authorized the employment of a new, heavier livre called the "livre poids de marc" that contained 9216 grains (489.506 g) and was subdivided in two different ways. Whenever such changes took place, of course, most or all of a unit´s submultiples or subdivisions were altered appreciably.

Hundreds of similar examples could be supplied for other European states. Central governments were actually encouraging metrological proliferation by such practices. If uniformity to national standards were the desired goal, such actions merely served to encourage the multiplicity of regional and local standards. Local populations grew accustomed to ignoring government directives.

Local Creations

Occasionally a common, local weight or measure would become so popular that it would gain either wide-spread local acceptance or even unit standardization. A measure of capacity for coal at La Rochelle called the baille, for instance, was eventually considered the equivalent of 1/80 muid. Originally it was any metal or wooden bucket used for carrying water. Most local creations, however, remained unfixed and unregulated. In England a trendle was any round or oval tub used for selling wax; a prickle, any wicker or willow basket for fruit; a costrel, any leather, wooden, or earthenware vessel for wine that was carried at a man´s side; and a coddus, any small bag for grain. The coste, fargot, and flin of France, and the balla of the Italian republics were similar in type and description.

Thankfully, some local measures never reached either status, but so many did that the disparity between state and local units grew increasingly more troublesome to the smooth functioning of business and commerce. Such local efforts oftentimes emulated the actions of central governments. The mistakes of London, Paris, and other European capitols radiated outwards to local metropolitan centers.

Town and Country

With the rapid growth of towns during the High Middle Ages, weights and measures in certain locales tended to be separated into different standards depending on whether they were employed within the original, early medieval, walled structures, or outside these walls in the expanding "suburbs" and areas beyond. This gave rise to scores of measures designated either as "intrà muros" or "extrà muros." An outstanding example of this was the denrée employed at Châlons-sur-Marne. As a measure of area for land, the "extrà muros" or denrée employed outside the walls was 5335 5/9 square Parisian pieds (5.630 a), while the intrà muros" standard of the town proper was 5555 5/9 square Parisian pieds (5.862 a). To further complicate the situation, the denrée was always reckoned equal to 1/9 arpent or journal whether inside or outside the town. Throughout the German States city measures usually carried the prefixes Stadt- or

Strassen-, while country measures were normally prefixed with Forst-, Marsch-, or Wald-. Among the most famous were the Stadtfaden of Lubeck, Strassenruthe of Dresden, Forstfaden of Lubeck, Marschruthe of Hamburg, and the Waldklafter, Waldmorgen, Waldruthe, and Waldschuh of Frankfurt.

Land and Sea

Similarly, some measuring units had different standards depending on whether they were used on land or on sea. The French lieue, for instance, was originally the distance that a man could traverse in one hour of ordinary walking and was used in Gaul before the Roman occupation. By the fifth century it was reckoned as 1500 Roman paces of 5 feet each or, in metric terms, 2.216 kilometers (1.47 m x 1500). By the end of the eighth century it had increased to 3 Roman miles or 4.411 kilometers. During the Later Middle Ages lieues of many different lengths were employed, most of them being between 2000 and 3000 toises, and the larger lengths were usually employed for sea distances. The "lieue marine" was eventually standardized at 3 milles de marine or 5555.62 meters, while the "lieue de Paris" for common road measurement contained 12,000 pieds or 3898.08 meters. In the German States the Seemeile was identical to the international nautical mile of 1.852 kilometers or 1.15 English miles, while the Postmeile was 25,400 Fuss (7.42 km) after 1818 in Hanover, 2000 Ruthen (7.53 km) from 1816 to 1872 in Berlin, and 2000 Strassenruthen (9.07 km) before 1858 and 1 1/80 Deutsche Meilen (7.50 km) from 1858 to 1871 in Dresden. The Wegstunde or "normal walking time" was traditionally 2.76 Meilen (ca. 4.44 km). English examples were the various lengths of the fathom and league. There were also numerous Italian examples as well. Measures always increased in size and distance once land was no longer in sight.

Products and Quantities

Product variations were the single most important source for metrological proliferation. If they were based on quantity measures, many of them would be indefinite in number, tale, or count, and would be based on an irregular assortment of human, animal, and other capabilities. Or they would refer to any convenient number of products for shipment by packtrain or other animal-pulled or ship-hauled devices. Usually enclosed within some canvas sack, bag, or bale, measures of this type were frequently employed. Examples of such units in England were the balet, bolt, fad, fadge, fardel, fardlet, fesse, flitch, packet, and trussell.[5] France had the ballot, quarteron, gerbe, telleron, and membrure, while in the Italian States one finds the balla, balletta, balloncielio, ballone, fagotto, mazzo, pezzo, torsa, and torsello.[6]

Even when these measures had standardized counts, capacities, or weights, the actual size depended on the characteristics or peculiarities of the product involved or on other factors. In England, for example, the bale for bolting cloth was 20 pieces; buckram, 60 pieces; fustian, 40 or 45 half-pieces; paper, 10 reams; pipes, 10 gross or 1440 in number; and thread, 100 bolts. A bunch of onions or garlic was 25 heads, while for glass it equaled 1/60 wey or 1/40 waw. A bind of eels consisted of 10 sticks or 250 in number, a binne of skins was 33 in number, a bottle of hay or straw was 7 pounds (ca. 3.175 kg), and a cage of quails was generally 28 dozen. A dicker of hides was 10; horseshoes and gloves, 10 pairs; and necklaces, 10 bundles, each bundle containing 10 necklaces. For grain a fatt contained 9 bushels (ca. 3.17 hl), but for bristles, 5 hundredweight (254.010 kg), coal, 1/4 chalder (ca. 3.17 hl), isinglass, 3 1/4 to 4 hundredweight (147.417 to 181.436 kg), unbound books, 4 bales equal to 1/2 maund, wire, 20 to 25 hundredweight (1016.040 to 1270.050 kg), and yarn, 220 or 221 bundles. A flock of piece goods was 40 in number or set; a glean of herrings was 1/15 rees or 25 in number; a gross of piece goods was 1728 in number; and a gwyde of eels was 10 sticks or 250 in number. A hundred for most products was 100 in number, but 120 for balks, barlings, boards, canvas, capravens, cattle, deals, eggs, faggots, herrings, lambskins, linen cloth, nails, oars, pins, poles, reeds, spars, staves, stockfish, stones, tile, and wainscoats. For cod, ling, saltfish, and haberdine, the hundred was 124 in number; for "hardfish," 160 in number; and for onions and garlic, 225 in number. A kip of lambskins was 30; of goatskins, 50. A load of osiers in Essex was 80 bolts and a maund of unbound books, 2 fatts or 8 bales or 40 reams. A pack of cloth was 10 pieces, but for flax, 240 pounds (108.862 kg), teasels, generally 9000 heads for kings and 20,000 heads for middlings, and yarn, 4 hundredweight or 480 pounds (217.724 kg). A mease of herrings was 500 to 630 in number, while a nest of piece goods consisted of 3 in number or sets. A quire of paper was 24 or 25 sheets equal to 1/20 ream; a rees of herrings was 15 gleans or 375 in number; a roll of parchment was 60 skins; a rook of beans in Yorkshire was 4 sheaves; a rope of onions and garlic was 15 heads; a roul of eels was 1500 in number; a sack of sheepskins in Scotland was 500 in number; a seron of almonds was 2 hundredweight (97.976 kg), of aniseed, 3 to 4 hundredweight (152.406 to 203.208 kg), and of castle-soap, 2 1/2 to 3 3/4 hundredweight (127.005 to 190.507 kg); a shock of piece goods was 60 in number; a skive of teasels in Southampton was generally 500 in number; a stick of eels was 25 in number; a stoke of dinnerware was 60 pieces; a timber for skins was 40 in number; and a waw of glass, 40 bunches. For barley, corn, and malt, a wey consisted of 40 bushels (ca. 14.09 hl), but for cheese it was 180 pounds (81.646 kg), flax, 182 pounds (82.553 kg), glass, 60 bunches or cases, salt, generally 42 bushels (ca. 14.80 hl), and lead, generally 182 pounds (82.553 kg), but occasionally 175 pounds (79.378 kg).

Hundreds of other units containing thousands of additional variations existed and they made the operation of regional and interregional commerce extremely difficult and complicated. It should appear evident how such a confusing condition contributed to constant fraudulent practices and to continuous misunderstandings in business transactions. The many medieval merchants´ manuals are a testimony to the enormous influx of such measures.

Numbers

It was customary in the Middle Ages to create additional weights and measures to fill ever expanding agricultural and industrial needs by dividing existing units into halves, thirds, and fourths, and, where such subdivisions were not practical or possible, into an irregular assortment of diminutives. The most important of these units were the demi (=half) series in France such as the demi-arpent, demi-aune, and the like.[7] Quarters of existing measures were also prolific and were usually prefixed with -quart.[8] In England such renderings were usually preceded by -farthing, -fer, -for, -fur, or -quart as in farthingdale, ferling, forpit, furendal, and quartern. Thirds in both countries were identified by the prefix -tierce or -ter.[9] In France and England more undefined diminutives existed than the aforementioned halves, thirds, and fourths, and they caused massive confusion, especially in local trade.[10] Throughout the German States the most common prefixes attached to hundreds of units were the -Achtel or -Achteling (1/8), -Drittel (1/3), -Halb or -Halbe (1/2), -Quart (1/4), and -Viertel (1/4).[11]

Multiple Standards, Names, and Usages

In the British Isles a major problem adding to metrological proliferation was that the same measuring units had different standards in England, Scotland, Ireland, and Wales. Chief among these units were the acre, pint, ell, furlong, gallon, mile, and tun; all immensely popular measures. As seen earlier in another context, those of Scotland were often double, triple, or quadruple the size of their English equivalents, while those of Ireland were usually slightly smaller.

In addition, occasionally one standard was based upon another. The English herring cran, for instance, was defined as the equivalent of 34 wine gallons (ca. 1.29 hl), but it was based upon a completely different capacity measure. A standard but buttomless 30 gallon herring barrel was heaped full and the barrel then lifted, leaving the herrings in a pile on the ground.

Some units had the distinction of being used for more than one measurement division. The French aissin was a measure of capacity for grain, a measure of volume for wood and plaster, and a measure of area for land. A corde was a measure of length for agricultural and forest lands, a measure of area for small garden plots, and a measure of volume for firewood. Such unit diversity was commonplace everywhere.

Further confusion was caused by the widespread tendency to designate a certain unit with more than one name. In France there were such multiple names as gros = treseau, absa = aune = canne, barrique = bussard, obole = maille, and denier = scrupule. In the British Isles there was the pint = jug = stoup, butt = pipe, hundredweight = quintal = cental, virgate = wista, rod = perch = pole = goad = verge, yard = perch, plowland = hide = sulung, and oxland = oxgang = oxgate. In Italy there was the grosso = dramma = quarro, degalatro = decimo = decina, danapeso = denaro, and cantaro = quintale = centaio = carara = centinaio.

Submultiples and Subdivisions

Even when there was only one standard and it did not fluctuate over time, it could have developed various methods of submultiple compilation. The best illustration of this was the fother used in England as a weight of 2100 pounds (952.539 kg) for lead. It was subdivided in four different ways: 30 fotmals of 70 pounds (31.751 kg) each, or 168 stone of 12.5 pounds (5.670 kg) each, or 175 stone of 12 pounds (5.443 kg) each, or 12 weys, each wey of 175 pounds (79.378 kg). Such confusion was rampant across the European continent.

Accounts

The English last was an example of a unit of account—a measure or weight that did not have a prototype standard because of its enormous size or dimension. Such measures were simply computational units used for record-keeping, transport, shipping, or determination of business profits. As capacity or volume measures they were reserved solely for wholesale shipments. In the case of the last, it had the following standards: ashes, barrel fish, butter, oatmeal, pitch, and soap, 12 barrels (ca. 17.76 hl); beer, 12 barrels (ca. 19.92 hl); bowstaves, 6 hundred; feathers, 1700 pounds (771.103 kg); flax, 6 hundred bonds; grains, 10 seams or 80 bushels (ca. 28.19 hl); gunpowder, 24 barrels or 2400 pounds (1088.616 kg); herrings, 12,000; hides, 20 dickers or 200 in number; potash, 12 barrels or 2688 pounds (1219.248 kg); raisins, 24 barrels or 24 hundredweight (1219.248 kg); salmon, 6 pipes or 504 gallons (ca. 19.08 hl); salt, 10 weys or 420 bushels (ca. 148.00 hl); tar, 12 barrels (ca. 14.28 hl); and wool, 12 sacks or 4368 pounds (1981.290 kg).

In France the laste was used for dry products in the import and export trade of Lille, 38 rasières for wheat and 40 rasières for oats, Marseille, 3 quintaux or 300 livres poids de table (12.238 dkg), and Montpellier, 2 milliers or 20 quintaux or 2000 livres (ca. 2000 kg).

In the German States the Last was as widespread as in the British Isles, but was used principally for bulkrating shipments in barrels, tuns, or vats of barley, beer, corn, herrings, hops, oats, peas, rye, wheat, and wine.

Wholesale and Retail Trade

In conjunction with units of account were measures that were reserved solely for wholesale shipments. In France the pièce for wine and brandy was employed specifically for regional and interregional wholesale transport over highways and waterways. Generally it did not have fixed dimensions, but represented any large cask, vat, tun, barrel, or other container loaded on wagons or aboard ships, and sometimes it merely referred to a given number of smaller receptacles. However, in the following mercantile centers, it normally had standard values: Bayonne, 80 veltes; Bordeaux, 50 veltes or approximately 105 English imperial gallons; Bourgogne, 110 pots of Lille; Cognac and La Rochelle, 75 to 90 veltes; Marseille, 700 to 1700 livres weight content; Montpellier, 1400 livres weight content or 5 to 5 1/2 barils; and Nantes, 29 veltes.

There were hundreds of units such as these and they usually escaped the detection of land and port authorities assigned to protect the financial interests of citizens, fairs, and markets.

Coinage, Wages, and Prices

It was customary in the Middle Ages to base agricultural area or superficial measures of land either on coinage standards or on units of income derived through production. In France the soudée represented either that amount of land required to produce an annual income of one sou (monetary), or that amount that could be acquired or rented for one sou, or that amount that was assessed on the tax rolls at one sou. The English librate was an amount of land worth one pound (monetary) a year. Its total acreage depended on local soil conditions and on the value of the pound, and it seems to have varied from several bovates or oxgangs (often four) to as much as 1/2 knight´s fee. The knight´s fee probably originated as an amount of land needed to support a knight and his family for a period of one year. In this sense, the knight´s fee was regarded as a unit of income for a fighting man just as the hide was probably a unit of income for a working man or serf. But as early as the thirteenth century, the knight´s fee was expressed as a land division containing a definite number of bovates, virgates, or hides and, even though there was little uniformity, the following were the most common: a knight´s fee of 4 hides, each hide containing 120 acres, or 480 acres (ca. 194.40 ha) in all; of 4 hides of 16 virgates, each virgate containing 4 farthingdales of 10 acres each, or 640 acres (ca. 259.20 ha) in all; of 5, 5 1/2, 6, 6 1/2, 8, 10, and 12 hides, no standard acreage established for the hide; of 12 hides totaling 600 acres (ca. 243.00 ha); and of 14, 16, 27, and 48 hides, no standard acreage established for the hide.

Europe abounded with such customary measures, and most of them were preserved only through oral tradition. The latter condition, of course, was a hallmark of medieval life.

Agriculture and Taxes

Measures were also based on food production and tax assessments. The toltrey was an English measure of capacity for salt (ca. 1400) containing 2 bushels (ca. 7.05 dkl). It was derived from *tolt*, toll, and *rey*, king, and was so named because it was a fixed toll on salt paid by the men of Malden to the Bishop of London. At Caithness in Scotland a boll of bear´s sowing was equal to approximately an acre (ca. 0.51 ha) and was used as a measure for the payment of rent. During the fourteenth century a measure of capacity for wine known as the caritas was standardized at Evesham, Abingdon, and Worcester at 3/4 gallon (ca. 2.84 l), 1 1/2 gallons (ca. 5.67 l), and 2 gallons (ca. 7.56 l) respectively. This caritas or "charity" originally was an allotment of wine given by an abbot to his monks over a certain period of time rather than a definite capacity measure. In France the fourée was merely the amount of arable land on which a crop of wheat could be harvested.

Such measures were common everywhere in feudal and manorial Europe, particularly so on ecclesiastical estates. All of these and similar measures varied greatly by individual, family, political connection, financial worth, and region of residence.

To complicate these variations and irregularities, capacity measures throughout Europe were either heaped, striked, or shallow. The heaped measure (*comble*, *coumble*, *cumulatus*) contained an amount of grain extending above its rim. The actual amount in excess of a level measure depended on the proportions of any local vessel, such as the English full, heap, ring, and fatt. A vessel in which the contents did not extend above the rim was a striked or level measure (*ras*, *rasa*, *rasyd*, *sine cumulo*, *stricke*, *stryke*, etc.), and in this category were the French rasière and the English strike, sleek, raser, and hoop. A shallow measure (cantel, cantell, grains sur bords) was one in which the contents did not reach the rim. Unfortunately, public and private employers usually demanded payments in heaped measures while they ordinarily rendered their compensations in shallow measures. Tremendous societal friction was caused in both manorial and nonmanorial Europe by such activities. Frequent riots and rebellions were aimed specifically at the eradication of these inequities.

Labor Functions and Time Allotments

Medieval land and product measures were also based customarily on work functions and time allotments. In Lincolnshire the brescia was employed for turf-cutting on the fens. It represented the amount of land that could presumably be dug annually by one man with a spade between May 1st and August 1st. In Herefordshire a math equaled approximately 1 acre (ca. 0.40 ha) or the amount of land that a man could mow in one day. Elsewhere in England a wash containing approximately 1 gallon (ca. 4.40 l) originated as the amount of oysters washed at one time, while a werkhop of grain of about 2 1/2 bushels (ca. 8.81 dkl) represented one day´s work in thrashing. In France the ouvrée for vineyards and hempfields was the extent that a worker (ouvrier) could plant, sow, spade, and the like in one day; the andain the space that a mower could cut on each side of him as he moved in a straight line from one end of a plot of land to the other; the bichetée the amount of land that one was able to seed with a bichet of wheat; the fauchée the space that a mower cutting hay could cover in one day; the soiture (derived ultimately from Latin sextura, sixth) the extent of land that a reaper was able to work in one-sixth of a work-week; the hommée (derived from homme, man) the amount of meadow-land that a man could cut with a scythe in one day; and the jour (from Middle French jour, time, hour, day) that was reckoned as the amount of land that a man could perform any work function on within an indefinite time span. Obviously, all of these units depended on the individuals involved, their specific work functions, and local topography. Regulation of such measures was an impossible task.

Production Spans

The production span or strength potential of one or more animals constituted still another method of establishing standards. A capacity measure for milk in Suffolk and Sessex known as the meal equaled the quantity taken from a cow at one milking. Along the Rhone river the ânée represented the stock of goods carried on the back of one ass; the actual amount depended on the size of the animal, the distance covered, and the condition of the roads. The benaton was originally any basket used to transport goods on the back of animals, generally asses, while the benne, used principally for the transport of coal, lime, and metal ores within mine shafts and on roadways, was usually a large, metal, rectangular receptacle fitted with wheels and pulled by a team of asses or mules.

Some linear measures were based on a specific number of steps or paces, on bodily feats or capacity, and on the range of the human voice. The mile, for instance, had a number of special variations prior to its standardization under Elizabeth I at 5280 feet (1.609 km). The medieval English mile was either 5000 feet (ca. 1.52 km) or 1000 paces of 5 feet each; 5000 feet or 8 furlongs of 125 paces each, the pace containing 5 feet; 6600 feet (ca. 2.01 km) or 10 furlongs of 220 feet each; or the Old English mile of 1500 paces, the pace varying in size from one region to another. In Wales the leap, a length of 6 feet 9 inches (2.057 m) after its standardization, was originally a normal jump for a "working man." A Welsh ridge, or space between the furrows of a plowed field, equaled 3 leaps or 20 1/4 feet (6.176 m), while a Scots fall of 6 ells (6.858 m) was the distance covered by dropping a rod, staff, pole, or stick on the ground. In France the houpée was the distance recorded between one man who remained stationary and shouted "houp" or "hop" and another man who walked down the road and stopped at that point where he could no longer hear the shouts.

No standardized system of weights and measures could possibly be formulated on such haphazard methodologies. It was unfortunate that such unit creations were so frequently encountered and influenced so negatively the most well-intentioned efforts of reform programs century after century.

Human Dimensions

Pre-dating the categories discussed above, of course, came measurement based on man's own body. Perhaps the oldest from the standpoint of time were linear measures based on the sizes or dimensions of human limbs and appendages. In England the digit—later standardized at 3/4 inch (1.905 cm)—was originally a finger's breadth, equal to 1/4 palm, 1/12 span, 1/16 foot, 1/24 cubit, 1/40 step, and 1/80 pace. The palm or hand's breadth was equal to 1/3 span or 1/6 cubit. Based on the foot of 12 inches, it was made equal to 3 inches (7.62 cm).[12] A span was equal to the distance from the tip of the thumb on the outstretched hand, and based on the foot it was made equal to 9 inches (2.286 dm).[13] The cubit was the distance from the elbow to the extremity of the middle finger, which was generally reckoned as 18 inches (4.572 dm), or 6 palms or 2 spans. A step was 1/2 pace or approximately 2 1/2 feet (ca. 0.76 m), while a pace equaled 2 steps or approximately 5 feet (ca. 1.52 m). Other body measurements were the shaftment of 6 inches (ca. 15.24 cm) or the distance from the tip of the extended thumb across the breadth of the palm; the nail, used principally for cloth, that represented the last two joints of the middle finger, equal to 1/2 finger, 1/4 span, and 1/8 cubit, and standardized at 2 1/4 inches (5.715 cm); the hand of 4 inches (10.16 cm); the finger for cloth equal to 2 nails or 1/2 span, and generally expressed as 4 1/2 inches (1.143 dm), and the fathom, the length of a man´s outstretched arms containing generally 6 feet (1.829 m).

Labor Needs

The final source of unit proliferation came from the many urban and rural craftsmen and tradesmen who created special uses for existing weights and measures to fulfill particular labor needs. The best example of this practice was the linear perch employed everywhere in the British Isles. Although the standard consisted of 16 1/2 feet or 5 1/2 yards (5.029 m), variations of 9, 9 1/3, 10, 11, 11 1/2, 12, 15, 16, 18, 18 1/4, 18 1/2, 18 3/4, 19 1/2, 20, 21, 22, 22 1/2, 24, 25, and 26 feet (2.743 to 7.925 m) were commonplace by the beginning of the Industrial Revolution. Perches of

16 1/2 feet and smaller were usually reserved for agricultural land measurements, while those larger than 16 1/2 feet were used by woodsmen in the forest regions and by town laborers engaged in draining, fencing, hedging, and walling operations.

In the German States there were far more such labor examples, among which the following were the most noteworthy: the Baufuss and Bauruthe (Bau = construction) of Altenburg, Aachen, Cologne, Erfurt, Gotha, and Leipzig, the Feldmesserruthe, Feldruthe, and Feldschuh of Dresden, Erfurt, Frankfurt, and Gotha, the Kettenfuss (Ketten = chain) of Hanover, the Landmassfuss, Landruthe, and Landschuh of Aachen and Dresden, the Tagewerk of Bayern, Emden, and Fulda, the Vermessungsfuss (Vermessung = surveyor) and Vermessungsruthe of Altenburg and Coburg, and the Werkfuss, Werkruthe, and Werkschuh of Coburg, Frankfurt, Gotha, and Leipzig.

Conclusion

As emphasized throughout, there are thousands of other examples that could be provided for each of these major sources of unit proliferation. Metrological diversity affected adversely all segments of society. This was the legacy of the past that concerned citizens, scientists, government ministries, and the bureaucracy had to contend with during the seventeenth and eighteenth centuries. The Scientific Revolution, expanding technology, industrial growth, and metrological reforms would provide the incentives to eliminate this enormous burden and create a workable, simplified system of weights and measures. European metrology was on the threshhold of a new era.

2
EARLY REVISION IN BRITAIN: THE SEVENTEENTH AND EARLY EIGHTEENTH CENTURIES

Medieval Bequests

When Elizabeth I died in 1603 the Tudor monarchy came to an end and the reign of the Stuarts began. There was little on the metrological horizon early in the seventeenth century to suggest that a new era in British weights and measures would be ushered in, but over the course of the next two centuries more significant metrological reform would occur than in the preceding thousand years. To be sure, there were many efforts made by pre-Stuart governments to streamline a constantly expanding metrology in the wake of a later medieval economic resurgency in agriculture, industry, business, and commerce. Monarchic decrees and parliamentary enactments defined many of the principal units of measurement in the hope of eliminating some of the tens of thousands of local and regional variations. Continuous scientific and technological progress, predominantly under the Tudors, led to the production of physical standards of ever increasing reliability and sophistication. Various governments were even able to trim somewhat the generally inefficient corps of administrators who had multiplied rapidly during the Middle Ages and Early Modern period and had acquired powers and duties that, especially in the political and economic realms, either rivalled or surpassed those bestowed on other representatives of the government bureaucracy. However, too many problems remained to plague even the most well-intentioned efforts.[1]

The major problem with weights and measures before the seventeenth century was the enormous disparity between the units employed by the central government and those employed locally and regionally. Beginning as early as the twelfth century, London tried to rectify this situation by issuing decrees and later promulgating legislative acts, by manufacturing and disseminating physical standards to prominent cities and markets, and by instructing and supporting a corps of officials who were supposed to inspect weights and measures, to verify their authenticity by comparing them to government prototype standards, and to enforce metrological laws. The goal was to align regional and local systems with those of the central government but, for the most part, these programs failed for three reasons.

First, metrological decrees and legislative acts were, on the whole, ineffective. The wording in most of them was extremely ambiguous. Standards were mentioned, but usually they were not defined or even identified; this was especially true of linear standards. The inch, for example, was described indiscriminately as the length of three round and dry barleycorns instead of relating it to the distance between two markings on some line bar. Capacity measures were not described in cubic inches, but rather by their weight content in certain dry products, liquids, and even by river water. The multiples and submultiples of weights were listed, but rarely were the systems upon which they were based described in grains or linked to some government standard. In several instances laws even promulgated different national standards for the same unit. Injunctions and prohibitions were repeated constantly producing a prodigious number of enactments which made a proper knowledge of weights and measures very difficult to obtain. Constant repetitions of centuries-old metrological formulas spelled non-compliance. Some laws provided exceptions, especially to the aristocracy and certain commercial interests, which, being a departure from the principle of uniformity, set precedents for still more exceptions. Some laws favored certain regions or provinces to the detriment of others. Frequent repeals, necessitated by hastily drawn decisions, seriously affected in a negative manner citizens' responses to new laws. Certain laws even acknowledged past failures but did not prescribe new approaches or stiffer penalties for non-compliance.[2]

Second, medieval physical standards were too few in number considering the many types of weights and measures authorized for legal use. There were no physical standards for many units employed by the central government and virtually none at all for sanctioned local or regional exceptions. Standards sent to cities and markets were always copies and, as such, they normally varied from the originals. But even the originals varied because they were constructed at different locations. For example, as many as five or six metropolitan centers supplied standards in England and Scotland whenever ordered by the Crown. Then errors were compounded when local artificers made copies from the copies they received. There was also continuous deterioration of these standards. Wooden standards decayed; those constructed from lead, iron, or bronze oxidized. State standards suffered from constant handling by officials, while local standards, posted to municipal or market walls, became impaired by weather conditions and general negligence.[3]

Finally, the overwhelming number of officials who were eventually entrusted with inspection, verification, and enforcement duties stifled standardization efforts. Townsmen acting individually or as members of ad hoc commissions, manorial lords and courts, church dignitaries, university administrators, urban magistrates, guildsmen, port officials, justices, sheriffs, coroners, government ministers, market personnel, and many others performed one or more metrological duties. There were too many overlapping jurisdictions and too many possible sources of fraud and corruption. Often their duties were poorly defined or they were not trained properly. In most cases one´s remuneration depended on the number and amount of fines levied, and this situation led to abuse. These negative aspects can be extended much further, but the central issue here is that the faults inherent in this and the other aspects of metrological control worked against standardization guidelines. By these practices the government was, in fact, contributing directly to metrological diversity and proliferation.[4]

These faults were not found solely in the British Isles, of course, since the metrologies of France, Italy, the German States, and other European areas show the same, and in some cases, even greater diversification. But the causes generally remained the same regardless of the country or culture involved.

The Era of Change

Confusion reigned in the metrological systems of the British Isles during the early decades of the seventeenth century. With such an overwhelming number and variety of weights and measures as showcased in preceding discussions, a severe strain was placed on many sectors of the economy. Merchants had to carry manuals that correlated the weights and measures of one country with another, of one region with another, of one city with another. Encyclopedias, dictionaries, and hand books abounded in the desperate hope of bringing some degree of comprehension to a potpourri of conflicting units and competing systems. Like the medieval serfs of an earlier era, the agricultural work force continually accused their employers or managers of manipulating the sizes of weights and measures to cheat them out of their due returns and to increase what they already considered to be unjust profits. In markets and trading centers an inordinate amount of time was wasted checking for fraudulent practices in the sizes and construction of weights and measures. Recourse to the local or regional standards usually accomplished little since they were either defective or had been altered to the benefit of a particular individual or group. Even referrals of corrupt practices to the metrological officer corps rendered little satisfaction in most cases since their duties were poorly defined or not defined at all, and there were so many individuals involved that a labyrinth of conflicting or overlapping territories and jurisdictions resulted making legal recompense a long, expensive, and laborious process.

Change was mandatory. Some change had occurred in earlier centuries. Nevertheless, to certain scientists, technicians, metrologists, and other concerned citizens after the Tudor era, the speed with which this change had occurred was far too slow. Decades, even centuries, had been necessary to bring about reforms, and then all too often the results were impaired by ineptly conceived and poorly worded legislation, defects of various magnitudes in the standards, and inexperience, excessive competition, or corruption among those chosen to enforce the reforms. To many people what was needed was a restructuring of certain aspects of this metrology so that its inherent weaknesses would not reactivate the same problems in the future. To others, such preventive proposals would never be successful since the fundamental structure of the entire system was considered at fault. Instead of repairing partially a metrology which could never be made totally operable, these reformers argued that it would be far wiser to overhaul thoroughly the old system, or even to dismantle it totally and replace it with an entirely new system. Some of them called for a strict decimal or other non-duodecimal scale for building unit proportions. Others, desirous of establishing new weights and measures standards, wanted a system based upon pendulums beating seconds at various latitudes or upon terrestrial measurements that had gained such international acclaim during the early stages of the Scientific Revolution. Still others concentrated their efforts solely on inspection, verification, and enforcement procedures, and believed that a simplified, workable form of the present system could be achieved if the contemporary administrative framework were eliminated and replaced by a smaller and more professionally qualified inspectorate. But regardless of their differences, all of them were motivated by one common belief—change could be speeded up. To them, it was imperative.

Radical change in British metrology, however, was not meant to be during the seventeenth and early decades of the eighteenth centuries. Instead, a policy of revision was adopted whose eventual failure at correcting most of the outstanding problems served as the impetus for the radical changes of the late eighteenth and early nineteenth centuries—changes that would usher in the imperial system of weights and measures in England, and, more importantly, the metric system in France. Before examining the revisionist era and the later sweeping alterations in these systems, it is necessary to describe briefly two aspects of the Scientific Revolution that related heavily to the eventual metrological reforms. Their impact would be slow in coming to the transformation of metrology in sharp contrast to their almost immediate impact on other fields such as horology, astronomy, and physics. But it would be the newly attained status of technological innovation throughout Europe and the newly created scientific societies that would change appreciably the evolutionary course of British and, for that matter, world metrology.

Technological Innovation

Of the hundreds of contributions of the Scientific Revolution of the sixteenth and seventeenth centuries, perhaps none was more important than the overthrow of the traditional acceptance of the absolute authority of ancient and medieval thought as it related to the operation of nature, of world systems, and of man's relationship therein. In previous eras, it amounted to rank heresy to question the authority of an Aristotle (384-322

B.C.), Pliny (A.D. 23-79), or other intellectual giant. When medieval man discovered something that contradicted established thought, he usually tried to modify his position so that it would be in alignment with prior accepted concepts. Or he simply operated within a dualistic framework that allowed for the existence of two truths or two levels of truth or two variations of the same truth. Hypotheses and experimentation—the hallmarks of the Scientific Revolution—were not even a part of the structure of medieval universities; rather those institutions strove to perpetuate and disseminate the accumulated lore of the past. To make matters worse, medieval philosophers and other educated persons were influenced all too often by magic, the occult, astrology, symbology, word and number associations, and an odd assortment of unnatural or supernatural beliefs. The Scientific Revolution helped to remove these roadblocks and encouraged critical assessment of old assumptions; by scrutinizing past beliefs and by postulating new hypotheses and testing them by the experimental method, they produced change. Nowhere was that change more physically evident than in technological innovation.

The invention of numerous scientific instruments and procedures during the seventeenth century (whose impact will be seen in following chapters) had one dramatic effect: for the first time in history the theoretician linked up with the technician or craftsman. For the former to test his hypotheses, he needed much more reliable and accurate measuring devices such as astrolabes, lenses, magnification glasses, navigational instruments, clocks, timepieces, scales, and weights and measures. The key here was the urgent necessity of obtaining higher standards of observation and measurement. Each new innovation had ramifications far beyond its own narrow scope of application. A chemist needing more accurate means of calibration or of weighing collaborated with metrological craftsmen to gain the new instrument or apparatus. Once introduced it had a profound impact on metrological evolution and upon other fields. The new attitude that questioned authority also enabled seventeenth-century man to adopt existing devices for novel uses. Regardless of the particular interplay, however, the merger between theoretician and technician led to dramatic technological innovation. In metrology improved standards and conceptualizations of new, more scientifically sophisticated systems emerged. A new age of weights and measures was eventually ushered in.

Scientific Societies

Much of the technological innovation of the seventeenth century would not have occurred if left simply to the occasional or chance meetings between theoreticians and technicians. It needed stimulus and support and these were provided by several European scientific societies. Developing rapidly during the 1600s, by the end of the century most serious scientists in Europe had become members and publication in the journals of these societies became the recognized manner of announcing the results of investigations. Various strata of society participated in addition to university-trained professors for among them were landed gentry, country physicians, clergymen, apothecaries, craftsmen, lawyers, and military personnel. Although some of the earlier groups, such as the Roman

Accademia dei Lincei of which Galileo Galilei (1564-1642) was a member, consisted of little more than occasional meetings held for discussion and criticism of private investigations, by the middle of the seventeenth century a number of societies were founded aimed at corporate scientific activity entailing research and analysis. One of the most important was the Florentine Accademia del Cimento which brought men together to work in common on important scientific tasks, chiefly on the experimental development of the ideas of Galileo, Evangelista Torrecelli (1608-47), and Vincenzo Viviani (1621-1703). The Cimento only lasted ten years (1657-67), but it was renowned for its technological apparatus operating in what is considered to be Europe's first physical laboratory. Their corporate researches were conducted on strictly regulated experimental lines and their conclusions were limited to the confines of observed evidence.

The two most important societies for the future development of metrology were the Royal Society of London and the Académie des Sciences of Paris. The Royal Society probably developed from an informal association of adherents of Francis Bacon's (1561-1626) experimental philosophy. They first met in London about 1645 to discuss natural philosophy; among them were mathematician John Wallis (1616-1703), John Wilkins (1614-72), whose interests centered on mechanical inventions, the physicians Jonathan Goddard (1617-75), George Ent (1604-89), and Christopher Merret (1614-95), and astronomer Samuel Foster (ca. 1600-52). A small chapter started at Oxford around 1649 when Wallis, Wilkins, and Goddard settled there to teach. In November of 1660 an organization for this Oxford branch was set up, temporary officers elected, rules established, and a tentative list of members determined. Acquiring its first charter in August, 1662, this organization made significant contributions until it began losing its most prominent members. It came to an end in 1690.

The London chapter, however, flourished due to such illustrious members as architect Christopher Wren (1632-1723), chemists Laurence Rooke (1622-62) and Robert Boyle (1627-91), Sir Robert Moray (d. 1673), the president before incorporation, mathematician Viscount William Brouncher (ca. 1620-84), the first president after incorporation, Goddard, Henry Oldenburg (ca. 1615-77), the Society´s secretary, and the diarist, John Evelyn (1620-1706). Holding their weekly meetings at Gresham College, the fellows incorporated by charter on July 15, 1662. The choice of Gresham College was a natural one since it had well-known professors of geometry, astronomy, and medicine lecturing there on a regular basis. Named the Royal Society of London for the Promotion of Natural Knowledge, and consisting of approximately 150 members by the end of the century, it was from its inception a private organization, very different from other major scientific societies. Unlike the French Academy, for example, the members received neither privilege nor pensions; they were granted no buildings, funds, equipment, fellowships, or laboratories; and they elected their own colleagues instead of having them appointed by a government ministry.

At their meetings they assigned special projects to individual members or to groups. At later meetings reports were given of the results. When a paper was read or an idea discussed, the matter usually led to a series of experiments conducted before the members either by the researcher or his team or by a fellow selected for that purpose by the Society. Casting their scientific net far and wide, their work covered such diverse fields as physics, chemistry, biology, medicine, and natural science, and they performed experiments with the pendulum, barometer, thermometer, and hygrometer. Many of the results of their experiments were published in the Society's Philosophical Transactions begun in 1665 by Oldenburg.

The French Academy had its origin in informal gatherings of philosophers and mathematicians in Paris around the middle of the seventeenth century. This group, which originally included such luminaries as René Descartes (1596-1650), Blaise Pascal (1623-62), Pierre Gassendi (1592-1655), and Pierre Fermat (1601-65), later expanded to add, among others, the mathematicians Pet. de Carcavi (d. 1684), Nicolas Frénicle (1600-61), Giles-Personne de Roberval (1602-75), and Gaspard Desargues (1593-1662), the astronomers Adrien Auzout (d. 1691) and Jean Picard (1620-83), and the physicist Edmond Mariotte (1620-84). By the latter decades of the century their ranks included Philippe de la Hire (1640-1718), Wilhelm Homberg (1652-1715), Nicholas Lemery (1645-1715), Gottfried Wilhelm Leibniz (1646-1716), Olaus Roemer (1644-1710), Joseph Tournefort (1656-1708), and Pierre Varignon (1654-1722). They met to discuss current scientific problems and to suggest new mathematical and experimental procedures. After 1657 the meetings were regulated under a formal constitution drafted by Samuel Sorbière (1615-70), the Academy's secretary. Foreign scholars such as Thomas Hobbes (1588-1679), Christian

Huygens (1629-95), Jan Swammerdam (1637-80), and the Danish anatomist Nicholas Steno (1638-86) were soon attracted, and at the suggestion of Charles Perrault (1628-1703), Jean Baptiste Colbert (1619-83) proposed to Louis XIV (r. 1643-1715) the establishment of an Academy to be funded by the government. Such a proposal was no novelty in France since many royal courts traditionally spent huge sums of money supporting dance groups, painters, musicians, architects, and others. The scientists simply constituted a further dimension of this practice.

Originally intending to concern itself with history and literature as well as the sciences, the humanistic disciplines were soon eliminated, and when the Academy held its first meeting on December 22, 1666 (slightly more than four years after the Royal Society), it was a gathering only of mathematicians, astronomers, physicists, natural philosophers, chemists, physicians, and others in the sciences. Supported by the government, the members were given salaries and later provided with excellent working facilities. The Academy obviously enjoyed economic advantages not shared by the private Royal Society, and with its greater resources it was able to attempt projects far beyond the limited financial capabilities of the English group. This would have a monumental impact on the future development of the French metric system. But because of this strong government backing, the Academy was treated sometimes as a scientific department of state and it occasionally had to face problems imposed on it by the king or his ministers.

The Parisian Academy met twice a week, usually on Wednesdays and

Saturdays in two rooms assigned to them in the Royal Library, and devoted their sessions to physics and mathematics alternately. The work of the astronomers Picard and Auzout was especially important to metrological history for they introduced the practice of using telescopes in conjunction with graduated circles for the precise measurement of angles. Picard also measured a meridian arc in northern France in order to compute the radius of the Earth. From 1669 the astronomical work of the Academy was carried out by Giovanni Domenico Cassini (1625-1712), while Jean Richer (d. 1696) made the discovery that a pendulum, in order to beat seconds, must be shorter at Cayenne than at Paris. Not only did the seminal work of these scientists mark the beginnings of speculation as to the exact shape of the Earth, but it provided metrologists with possibilities for a natural physical standard which led eventually to the establishment of the metric system.

Unlike the Royal Society, reports of the proceedings during the early years of the Academy were never published, but it appears that they were very formal. Much smaller in membership than its London counterpart due, in part, to its government sponsorship, its meetings consisted of papers delivered by individuals or groups detailing the results of projects chosen by the members or assigned by the Academy. They examined new inventions and certified those approved on behalf of the state. Among their numerous contributions to technological innovation, the most noteworthy were in astronomy for they developed the telescope of very long focal length to its useful limit, applied the telescope to measuring instruments, and perfected the use of the telescope micrometer.

Astronomical Observatories

Two vitally important institutions that developed during this period as offshoots of the Royal Society and the Parisian Academy were the Greenwich and Paris Observatories. Their impact on the future metrologies of England and France was highly significant. The early history of the Greenwich Observatory is tied to the work of John Flamsteed (1646-1719), the first astronomer to hold office there. Born near Derby, he spent most of his life in the study of mathematics and astronomy. After having won the admiration of Oldenburg and Sir Jonas Moore (1617-79), the Surveyor of the Ordnance, the former encouraged his intellectual pursuits while the latter furnished him with scientific instruments necessary to set up an observatory at Derby. He later graduated from Cambridge University with a master´s degree. In 1675, Moore called him to London to take charge of an observatory that he established at Chelsea College, a property of the Royal Society. During this same period Moore had Flamsteed made a member of a commission, along with himself, Brouncher, Wren, and Robert Hooke (1635-1703), appointed to consider a proposal of a French lord named St. Pierre for obtaining longitude at sea by a method involving the precise determination of the Moon´s place among the stars. The idea had been promulgated earlier that the Moon, which goes around the sky in a month, could serve as a clock giving Greenwich time. If astronomers produced an almanac predicting the exact position of the Moon vis-a-vis the stars, navigators would possess a means of determining Greenwich time. Then by substracting such figures from local ships´ time the longitude could be determined. After observation proved this to be infeasible, Charles II (r. 1660-85), immensely impressed by Flamsteed, gave him the position of "Astronomical Observor." Soon thereafter, Wren´s suggestion that the site known as Greenwich Hill would make an ideal observatory was accepted, and in June, 1675, the history of the Royal Observatory began. On the site was a large tower and the Observatory was built on its foundations. In time astronomers Flamsteed, Edmond Halley (1656-1742), James Bradley (1692-1762), and Nevil Maskelyne (1732-1811) carried on observations here of the sun, moon, planets, and bright stars, and added substantially to the accumulated knowledge of the positions and movements of these bodies. Gradually, with the assistance of mathematicians and opticians, they improved numerous scientific instruments. Their contributions, among those of others working here, would eventually revolutionize metrology.

The Paris Observatory was an offshoot of the Academy. The foundations were laid in 1667 and the building was completed in 1672. Ironically, the London Society was begun more than four years before the Paris Academy, but the Parisian Observatory initiated its activities three years before Greenwich. The first astronomers to work here were Huygens, Picard, Auzout, and Cassini. Their most important contributions were in the application of the telescope to older precision instruments, and in the use and application of the pendulum clock. Picard obtained an improved value for the length of a degree of latitude on the earth´s surface in the area of Paris. For this purpose he measured an arc during the years 1669-70 extending from a point near Amiens to a point near Paris, and astronomically determined the difference of latitude at its extremities. To increase the accuracy of his survey, he connected the arc with a carefully determined baseline by triangulation—a method first proposed and perfected by the Dutch mathematician Willebrord Snell (1580-1626) from 1615 to 1617.

The work of the Royal Society was independently motivated; the successes of the Parisian Academy were sponsored by the government. It is again ironic that their roles were reversed when it came to the observatories. Unlike Greenwich, there was no central authority or fixed program of work at Paris. Each observer worked on whatever pleased him; very often out of his own residence. Both of these institutions, however, made substantial contributions upon which metrologists could base new systems of weights and measures in the later years of the eighteenth century.[5]

Revisionist Programs

As mentioned, the impact of technological innovation and of the contributions derived from the scientific societies and observatories would not influence metrology appreciably until the later eighteenth century, but some benefits would be felt during the revisionist era. The metrological policy of British governments during this period had five major aims. All of them represented marked improvements over past programs, and all of them were undertaken to achieve the goal of a simplified, integrated, and coherent system of weights and measures.

First, beginning early under the Stuarts, the government decided not only to continue the Tudor policy of constructing scientifically reliable and esthetically elegant physical standards, but to refine them and include standards for weights and measures not previously considered. They were to be manufactured at the Exchequer and at no other location. The aim was to have state standards for all of the principal units authorized by the government, to distribute accurate copies of them to the major population centers, and to destroy any in the existing arsenal that proved defective after testing.

In line with this was a second aim of incorporating any scientific or technological innovations that would increase the standards' precision. The best example of such a bold departure from past policy was the introduction of linear standards whose exact size was determined not by their overall length but by calibrating the distance between delicately etched line markings or brass insert pins located near the ends of the bars.

In regard to weights and measures legislation, revisionist policy attempted to list all of the acceptable national units, to describe them in precise language, to designate the exact standards on which they were based, and to increase the fines levied for non-compliance to serve as a strong deterrent. The ultimate purpose, of course, was the elimination of local and regional variations and systems. As will be seen, of the five programs involved here, this was the least successful.

Fourth, all future attempts to alter or change certain segments of the metrology of the British Isles would be entrusted initially to competent commissions whose members must be distinguished science, business, and government leaders. They were to study thoroughly the issue at hand, draw up a list of recommendations, and test any applicable physical device or standard manufactured in accordance with their directives at carefully controlled and regulated trials. This latter stipulation emulated the procedure of the scientific societies where members performed experiments at meetings to verify the results of their research. Only after all of these stages had been completed successfully would the government begin its deliberations and undertake to legislate change.

Finally, throughout this era governments strove to streamline the burgeoning metrological officer corps swollen to enormous size and reaping financial benefits and social prestige derived over the course of many centuries. The hoped-for result was a small, closely supervised corps of metrological specialists. As in the case of legislation, this proved an almost insurmountable task at this early date in the face of entrenched special interest groups.

The overall goals of this revisionist era enjoyed only marginal success, but out of the failures resulted the inspiration for the conceptualization and implementation of a radically new program in the later eighteenth century.

Refinement of Physical Standards

The most successful program of the revisionist era dealt with physical standards. Under the Stuart and later monarchs, weights and measures standards were refined further and several important issues supplemented the arsenal bequeathed by the Tudors. During the reign of James I (1603-25), the Exchequer delivered to the Founders´ Company a set of avoirdupois shield-shaped weights to enable the masters to perform properly their newly acquired task of sizing and marking brass weights used, made, or sold in London and its metropolitan area. The set was well designed and handsomely embossed. Frequent use, however, has severely damaged many of the designs as well as the surface edges. These weights were hung by a staple attached to the top of the shield´s face, a common design even in the Middle Ages (see Figure 1). In order to provide a complete set of standards for the two authorized weight systems of most of the British Isles, the Founders received from the Exchequer in 1684 a pile of flat bronze weights from 256 ounces to 1/16 ounce. All of them were stamped with the Exchequer seal for authenticity, marked with a crown over CR (for King Charles II), dated, and engraved with the initials of the maker. The latter practice became commonplace in order to determine ineffectiveness among certain craftsmen should the standards prove defective in future trials. A set of bell-shaped brass avoirdupois weights, sealed and marked in a similar fashion, was also supplied.[6]

Thirteen years before the latter grant, the Exchequer sent a standard yard, based upon an Elizabethan model, to the London Clockmakers´ Company. Horologists had recently separated from other crafts such as lockmakers, and in their newly earned status they needed standards for precision measurements. This standard of 1671 is a brass, octagonal rod, nearly 1/2 inch thick, and stamped with the Exchequer seal and the crown over CR. Unlike any previous linear measure, however, the length of this yard is expressed by the distance between two upright pins, or small cheeks, attached near both ends of the rod. A milestone had been reached for such standards would be very common by the eighteenth century, and this design would replace earlier bars whose size was determined simply by their overall length.[7] Such older standards were never effective since their ends could be shortened or damaged through careless handling, the effects of temperature, and the like.

Other standards dating from the Stuart period and preserved at the London Science Museum are a folding yard of 1634 (see Figure 2), a bronze gallon of the same date (see Figure 3), a grain quart of Charles I (r. 1625-49) (see Figure 4), a half-tod weight (14 pounds) of Charles I similar in design to Figure 1, a bronze wine pottle (2 quarts) of 1641 (see Figure 5), and a Winchester bushel of Charles II similar in design to Figure 7. The folding yard—the only one of its kind ever made by the government—is crudely designed and amateurishly constructed. If technological innovation were the asserted goal, this standard emitted too much potential error. The joints themselves raised the margin of error substantially. The first 12 inches are marked out in hastily drawn Roman numerals. The names Fovlke & Cartwright are printed in large Gothic letters on one side, while both ends carry a crown over C. The gallon of 1634 is embossed with the names of these same makers and contains a crown over CR. The grain quart is undated, of wooden construction with hooped brass edges at the top and bottom, and stamped with a crown over C. This also was a foolish choice since the expansion and contraction rates of wood and brass are very different. Also the "shelf-life" of wooden standards always proved to be dismally short in comparison with those constructed entirely of brass. The fourteen pound, or half-tod, weight is in very poor condition with badly damaged edges and with embossed reliefs that are almost unrecognizable. There is a return in this particular manufacture to some of the basic reliefs found on the weights of Henry VII (r. 1485-1509). The most expertly designed standards are the bronze wine pottle and the Winchester bushel. The former, conical in shape, is boldly stamped with a crown over CR, the royal coat-of-arms, the date of 1641, and the name of the official responsible for its sizing. The latter, although badly marked and unevenly constructed, is important since it is the only bushel surviving from the period between Elizabeth's death in 1603 and William III's accession in 1689.

As these examples show, there was a rather prolific production of standards throughout the seventeenth century. It must be emphasized, however, that some of them were duplications or slight alterations of previously established Tudor weights and measures. After the Tudor standards, and after the special standards mentioned above had been issued, no basic alterations would be forthcoming until the construction and legalization of the wine gallon of 1707, the fifth year of the reign of Queen Anne (1702-14). This is true even for the standards of William III (r. 1689-1702), four of which—constructed between 1689 and 1695—are preserved at the Science Museum, London.[8]

The Exchequer wine gallon is one of the most famous standards in British history. Prior to Anne, no wine standard had ever been constructed even though as early as Magna Carta (1215) an appeal had been made for a common measure for wine, and in Edward I's (r. 1272-1307) Tractatus (1303) the gallon for wine had been described as a vessel containing 8 tower pounds of wheat (43,200 troy grains or 2799.36 g). Repeatedly throughout the later medieval and early modern periods unit standards for wine were defined in parliamentary legislation with specific reference to so many wine gallons, but no more precise wording than that found in the Tractatus was ever provided. This uncertainty led, in 1700, to a governmental dispute concerning its legal contents and the amount of excise duty payable thereon. The Court of the Exchequer was unable to resolve the problem and referred it instead to a parliamentary commission, assembled in accordance with revisionist policy. This specially selected group quickly discovered that it had been customary, though no sufficient authority could be found for it anywhere, to use a wine gallon of 231 cubic inches. No such gallon had ever been officially promulgated by statute or royal decree, or had been recommended by government ministry or royal commission. The only legal standards for the gallon then applicable were the Exchequer grain gallon of 272 cubic inches and the Exchequer ale gallon of 282 cubic inches. To remedy this awkward and confusing situation, the customary wine gallon was legalized in 1706 and the hogshead of 14,553 cubic inches (63 gallons of 231 cubic inches each), the pipe of 29,106 cubic inches (126 gallons of 231 cubic inches each), and the tun of 58,212 cubic inches (252 gallons of 231 cubic inches each) were defined in terms of this new standard. In a statute Parliament decreed that any round vessel, commonly called a cylinder, having an even bottom and being 7 inches in diameter throughout and 6 inches deep from the top of the inside to the bottom, or any vessel, of whatever shape, containing exactly 231 cubic inches, was a legally acceptable gallon for wine. Although this was the first time in British history that a legislative enactment described a capacity measure in cubic inches, it was absurd for Parliament not to standardize officially its actual physical appearance. Leaving its shape to individual choice simply invited circumvention of the law.

The Exchequer standard gallon made in accordance with this act is still in fine condition and is kept at the Science Museum in London (see Figure 6). It is of heavy bronze, with a delicately designed handle on one side and a magnificently embossed crown over AR opposite it. Positioned equidistantly between them are oval scrolls of leaves. Within the latter, in bold script, are the words—Wine Gallon 1707. The rim is stamped with verification marks and with an alternating pattern of the crown over AR and the Winchester portcullis.[9] In a special examination conducted by the Standards Department during 1931-32, the actual capacity of this gallon was found to be 230.824 cubic inches, a figure that would be found grossly inadequate by today's standards, but remarkably accurate by early eighteenth-century expectations.

The precedent of legalizing the customary wine gallon was followed soon after by the legalization of the customary coal bushel. This measure held considerably more than the Winchester grain bushel, defined by the act of 1701 as a round measure, with a plain and even bottom, 18 1/2 inches wide throughout, 8 inches deep, and conforming to the standard in the Exchequer. Such a definition would render a bushel of 2150.42 cubic inches (35.238 l). The actual capacity of this and other Exchequer standards, however, did not conform to their statutory definitions as the examinations by the Standards Department in the nineteenth and twentieth centuries testify. A bushel of Elizabethan vintage actually measured 2148.28 cubic inches, whereas a standard of Henry VII, upon which Elizabeth´s was based, contained only 2144.8 cubic inches. To further complicate the situation, the Winchester gallon, one-eighth part of the bushel, was described prior to 1701 as having a capacity of 272 1/4 cubic inches. The Exchequer grain gallons of Elizabeth and Henry VII, however, were smaller, being 268.97 and 268.43 cubic inches respectively. Consequently, if eight of the statutory gallons of 272 1/4 cubic inches were added together, the result would be 2178 cubic inches or a bushel considerably larger than the statutory definition or the contemporary physical standards; another example of statutory definitions not coinciding with actual physical standards. Failing to recognize the differences among these standards or to rectify the discrepancies in the statutory definitions, Parliament enacted in 1713 that, in accordance with the customary practice in the port of London, the coal bushel must be round, with a plain and even bottom, and measure 19 1/2 inches from outside to outside. Its content was made equal to one

Winchester bushel (as defined by the act of 1701) and one quart of water. Again, such a contorted and awkward definition made its actual or practical application virtually meaningless.

In chapter 5 of the 1713 enactment, Parliament ordered the Lord High Treasurer, or any three or more of the Treasury Commissioners, to supervise the construction of a brass standard conforming to the definition above. After it was built, the Exchequer would seal it and provide for its safekeeping. This was not accomplished, and based on the above description, it should not appear surprizing that its actual construction was never undertaken. There is no record of any coal bushel before the reign of George II (1727-60). Preserved at the London Science Museum is one dating from 1730. It is made of bell-metal, inscribed "Ano [sic] Regni Georgii Secundi Regis Quarto" on one side and "The Coal Bushel" on the other. It was issued shortly after the act of 1730 had established additional regulations for the coal trade and had reiterated some of the provisions of Anne´s statute of 1713 (see Figure 7).

Legislative Aims

If the moderately successful physical standards program was ultimately tarnished due to the faults delineated above, the legislative efforts of both England and Scotland suffered to an even greater degree. This situation should not have happened since no single era in British metrological history witnessed such a profusion of enactments aimed at uniformity and standardization—43 separate weights and measures statutes issued between the years 1607 and 1758. This represented the most massive

Figure 1. Avoirdupois 28 Pound Woolweight (1614) of James I. (Lent to the Science Museum, London, by Mr. Owen Hugh Smith)

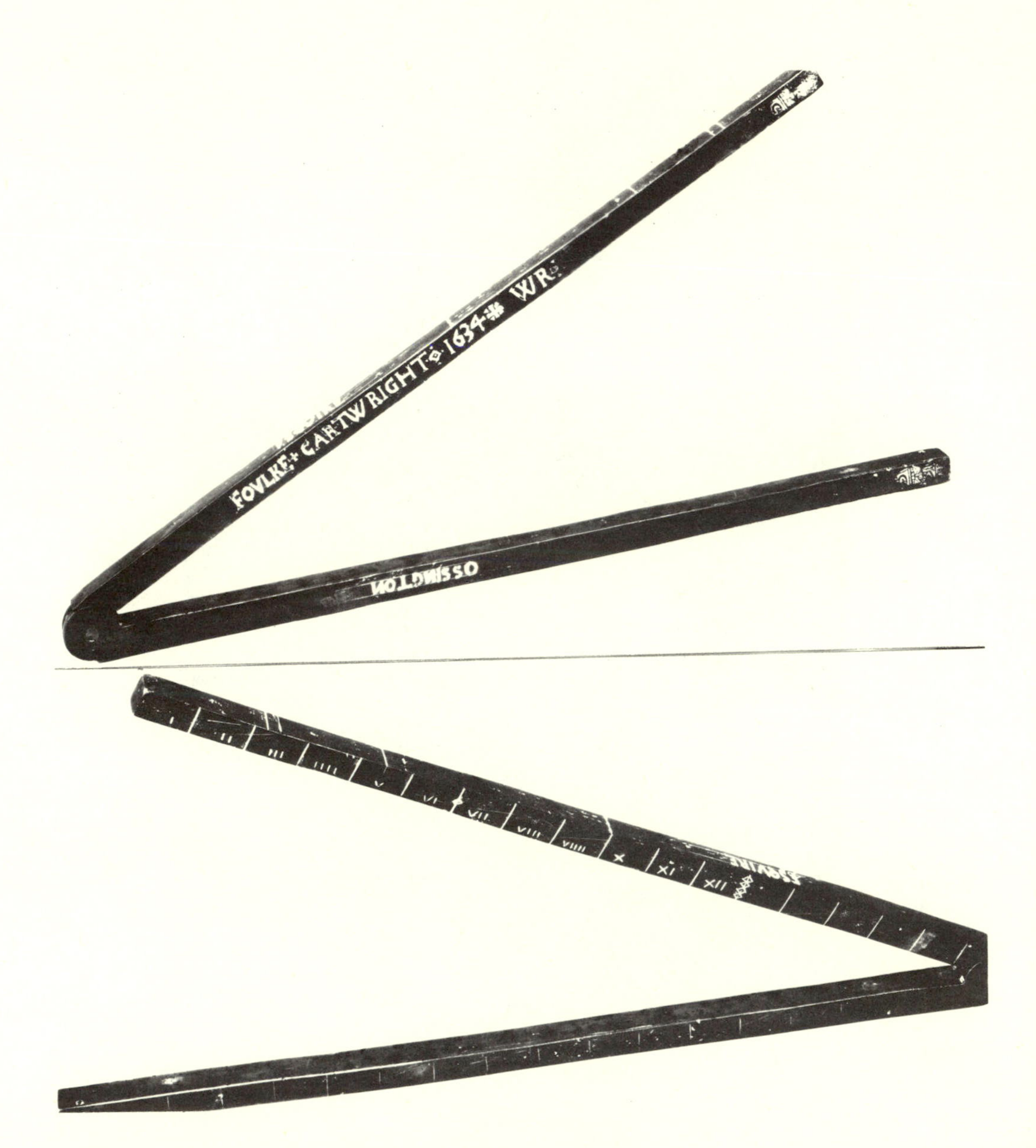

Figure 2. Folding Yard of 1634.
(Photo. Science Museum, London)

Figure 3. Bronze Gallon of 1634.
(Photo. Science Museum, London)

Figure 4. Grain Quart of Charles I.
(Photo. Science Museum, London)

Figure 5. Bronze Wine Bottle (1641) of Charles I.
(Photo. Science Museum, London)

Figure 6. Queen Anne Exchequer Wine Gallon of 1707.
(Crown Copyright. Science Museum, London)

Figure 7. Exchequer Coal Bushel (1730) of George II.
(Crown Copyright. Science Museum, London)

assault ever undertaken on metrological non-conformity. But aside from several positive steps to be discussed later, the overwhelming impact of this legislation was negative. There were many reasons for its eventual failure.

First, some acts, either in whole or in part, were simply complaints by the government of the continuing use of illegal weights and measures and of the widespread acceptance by the general public of fraudulent practices. These acts of 1607, 1617, 1640, 1661, 1681, and 1715 did not name the illegal units, did not specify what the practices were, and did not state what would happen if observance of the law were ignored. Further, to issue a rallying-cry for the use of statute measures (such as the acts of 1618 and 1661) which were never named, and which often could not be found in the law since they were never described, was counter-productive. Even naming the measures but not defining their dimensions as in the acts of 1681 and 1696, or pleading for striked in lieu of heaped measures as appeared in 1609 and 1670 without specifying exact cubic capacities, produced the same results. Constant repetitions of defective statutes not only signified non-compliance to all concerned, but they put the government on the defensive. Each time a statute was worded ineptly, or failed to list proper rules to be followed, or lamented non-observance, it merely reinforced popular defiance. What was needed was more diligent enforcement of fines and punishments.

To complicate matters, the statutes of 1617, 1618, 1621, 1625, 1630, and 1702 added a further dimension to this dilemma by complaining that former legislation was not observed and was not being enforced. Usually such statutes named the earlier acts, but if they were checked out, one would still be in a quandary due to their ambivalent nature. What was to be observed and who was supposed to do the enforcing? Then, when acts such as those in 1621 and 1625 threatened "grave punishment," but did not say how or by what method, the government was compounding confusion, inviting resistance, and falling deeper into the quagmire of persistent and defiant localism.

Other examples of legislative ineptitude along these lines can be found in the practice of constantly requesting various members of the officer corps to perform their duties of inspection, verification, and enforcement. Some acts outlined procedures to be followed in stamping or marking tested weights and measures and in the allowable fees to be collected for professional services. But enactments in 1609, 1617, and 1670, for example, simply pleaded for the observance of job functions. Since many of these people were appointed locally or regionally, the government did not know what those functions actually entailed. If one accepted a metrological function only as a financial sinecure to enhance social prestige, it was foolhardy for the government to insist upon non-existent national job descriptions. Even when commissions were appointed to correct abuses as in the acts of 1617, 1656, and 1661, too many important duties were left to individual or group determination rather than following strict government procedural guidelines.

Two particular onerous practices inherited from the Middle Ages continued throughout this period—the almost incessant repetition of certain metrological phrases that had long since become cliches and the insistance of depositing national standards for different weights and measures at various locations. The best illustration of the first was an act of 1640 that called for "one weight and one measure" to be employed throughout the kingdom. It is ludicrous that this phrase, originally drafted in a metrological decree of Richard the Lion Heart (r. 1189-99) in 1189, still found its way into statutory law almost five centuries later even though its ambiguity must have been apparent to lawmakers. Originally intended as a battlecry for standardization and uniformity in a far less complex economic and technological period, its use in the seventeenth century when tens of thousands of local variations had been added to British metrology was hopelessly archaic. An example of the second was the Scottish practice, as seen in the acts of 1618 and 1663, of depositing legal statutory physical weight standards at Lanerk, the standard ell at Edinburgh, and the standard pint at Stirling. Obviously, local interests and jealousies were the prime motivators here, but the overall effectiveness of well-intentioned legislative efforts was thwarted. In an age of backwardness in communication and transportation facilities compared to later centuries, governments were building obstacles to national metrological unification. Complete sets of standards should have been deposited in a predetermined number of sites selected on the basis of population density, scientific, technological, or economic importance, ease of access, and other vital factors.

Legislative Aims

Even though English and Scottish legislation was saddled with all of these problems, there were four areas in which positive results were obtained. First, and of critical importance for the future of British metrological unification, was the fact that in 1707 England and Scotland became the Kingdom of Great Britain by the Act of Union. In one spectacular move all Scottish weights and measures were declared illegal. One Parliament was formed in London, and the coinage, weights, and measures of England became the standards. Even though Scotland's metrology would linger on for more than a century (just as traditional English units today persist despite the Metric Conversion Act of 1975), all weights and measures legislation after 1707 was intended for the kingdom and not just for England. Scotland was supplied with English standards from the Exchequer and had to abide by English metrological law. A monumental stride toward eventual simplicity and uniformity was achieved.

Further, by the later seventeenth century parliamentarians began the practice of detailing specific monetary fines for metrological infractions. Initially amounting to a little as 5 pence, by the second and third decades of the eighteenth century, illegal practices warranted almost previously unheard of amounts from 50 to 100 pounds. In the statutes of 1681, 1696, 1697, 1702, 1707, 1709, 1717, and 1729, the fines rose steadily. The government was finally recognizing the obvious deterrent. Two statutes in 1696 even threatened imprisonment or incarceration for non-compliance. Another milestone had been reached. These were the first instances in British history where legislation threatened punishment by incarceration for a definite period for infractions of weights and measures law.

Finally, no fewer than 31 statutes, in whole or in part, promulgated definitions for weights and measures.[10] Among them, according to various product specifications, were the barrel, boll, bushel, cart-load, chalder, ell, firkin, firlot, foot, gallon, half-peck, hogshead, inch, keel, kilderkin, mile, peck, pint, pipe, pot, quart, quarter-peck, stone, tun, wagon-load, and yard. An act of 1620 was the first to describe the foot as the length of 12 inches, even though a statute in 1685 still continued to define the inch as the length of 3 barleycorns set lengthwise. The sheer number unfortunately produced a difficult situation. If only one statute after 1707 had been devoted to a complete listing of the acceptable units and a precise catalogue of those local and regional units deemed illegal, the government would have solved its unit dilemma once and for all.

Metrological Officers

Failing almost to the same extent as the legislative efforts was the last revisionist policy—the elimination of the thousands of inept, inexperienced, and unqualified weights and measures officials. Vested interests were too deeply entrenched, and the legislation never eliminated specifically any positions. The laws usually called simply for proper performance of metrological functions without describing them in detail. However, in several cases, there were slight improvements. For instance, special provisions for the inspection of weights and balances of pewter or brass were extended originally to the Pewterers' Company of London in two statutes of 1503 and 1512. In addition to powers similar in nature to those given other crafts in earlier eras, the master and wardens of this guild were permitted to appoint inspectors to fulfill the obligations of their charter. Their functions were spelled out in detail by the government in the seventeenth century. Likewise, an act of James I in 1611 authorized the Plumbers´ Company to "search, correct, reform, amend, assay, and try" all lead weights in London, its suburbs, and within seven miles of the metropolitan area.[11] They could enter any house or business establishment to conduct their tests. The Company had an office at the Guildhall and a member permanently stationed there to size and seal weights. He also had custody of the scales used in the examinations. As late as 1757 there is evidence that such operations were still being conducted in a professional manner. However, with the discontinuance of the use of pewter and lead as materials for physical standards manufacture, the metrological privileges of both these guilds gradually disappeared by the end of the eighteenth century. This is another rarity since most groups never relinquished any privileges.

The last of the major London craft guilds to be given metrological duties, and the most important from the standpoint of the standards as seen earlier, was the Founders´ Company. By a charter of 1614, they received permission to size and mark all avoirdupois weights sold or used in London, its suburbs, and within three miles of the metropolitan area. They had authority to search out defective brass weights and to destroy them. Until 1889 all brass weights sold or used in London were required to be stamped by their members in collaboration with an official stamper appointed by the

Corporation of London. These were positive steps undertaken by government initiative. They should have been extended to all groups or individuals performing metrological duties.

Even these meager results, however, were not realized in Scotland. Here justices of the peace received metrological functions in 1617 when an act of James VI (James I after 1603) declared that they were to conduct examinations of weights and measures in every burgh. Prior to each assize, town magistrates and guild deans had to certify how many and what types of weights and measures existed in their territories. When the justices received this information, the examinations commenced. After comparing them with the authorized standards, those found unlawful or defective were declared unfit for commercial use, and the king's council was promptly notified so that appropriate action could be taken. No weights and measures thereafter were legal if they were not listed on the justices' rolls and if they had not been previously examined. Nothing is said pertaining to verification methods or marks.

One year later an act reaffirmed these provisions while issuing a progress report on a recent commission entrusted with the task of establishing accurate standards. Since the commissioners found a continuing use of non-standard weights and measures, especially in the shires of Lanerk, Wigtown, Dumfries, Roxburgh, and Berwick, sheriffs in each of these five shires had to notify the justices and other officers to assemble in the major borough of each shire. Within 20 days after the council had set a proper time to begin operations, they had to collect all weights and measures again and reexamine them against the standards. All variations had to be registered so that proper price adjustments could be made by buyers in future commercial transactions. In this and the former act, it is extraordinary that defective measures were not ordered destroyed if they could not be rectified. Also, nowhere in the first act is there a clue as to what appropriate action the king´s council would take in the event that defective weights and measures were brought to their attention. Here, merchants and others could continue using them as long as every person concerned knew about the defects and made the necessary adjustments. Justices could not order them confiscated or burned, only record them in the proper manner. It is impossible to determine what purpose this served or how the government could use this information.

This being the case, it is not surprizing that another complaint charging diversity in weights and measures was made by the Scottish Parliament in 1661. After impaneling another commission to study the problem, the act commands the members to convene at an appropriate time and place to determine how such diversity could be eradicated and uniformity finally achieved.[12] When their decision was made, a copy of their deliberations had to be sent to the justices who would carry out their instructions. Additional commissions called for the same purpose followed and they attest to the program´s failure. Parliamentary and other sources never infer whether the commissioners were unable to formulate effective proposals or whether the justices were lax in carrying them out. Another statute in 1693 ordered justices "to take tryall...and to put this Act and other Acts thereby ratifyed to Execution." A final statute in 1696 dealing with the selling of meal by weight makes the same plea. Nothing was achieved. It took the Act of Union in 1707 to solve Scotland's metrological impass.

3

LATER REVISION IN BRITAIN: THE LATE EIGHTEENTH AND EARLY NINETEENTH CENTURIES

In 1758 a committee of the House of Commons appointed to determine the original standards of British weights and measures stated that

> Few people were heretofore able to make proper Measures or Weights: Standards were made and destroyed as defective, that others no less so might supply their Places; and the Unskilfulness of the Artificers, joined to the Ignorance of those who were to size and check the Weights and Measures in use, occasioned a great Number of different [standards]...to be dispersed throughout the Kingdom, which were all deemed legal, and yet disagreed. This Evil was encreased by the Means used to prevent it, _viz_. the sending [of] what were throught Standard[s] to all the Cities...for these being neither skilfully made, nor exactly conformable to the true Standards, of which they were strictly only Copies, became themselves again the Standards for other Copies, made and used in different Countries [districts] by Artificers, yet less skilful than the first Makers. These [local] Standards...were resorted to when those in the...Exchequer were neglected.... Thus every Erroe was multiplied, till the Variety...rendered it

> difficult to know what was the Standard, and impossible to apply any adequate Remedy.[1]

In the early eighteenth century a member of the Royal Astronomical Society echoed similar sentiments.

> Notwithstanding the propriety and advantage of having one uniform standard of measure, with which all other standards may, in case of any dispute or doubt, be compared, it was not till [recently]...that this country possessed any uniform standard to which an appeal could be made, even for ordinary purposes.... Although it had been declared as far back as...*Magna Charta*, that there should be an uniformity of weights and measures throughout the realm; yet the legislature has not been able...effectually to accomplish this object: one principal cause...has been the loose manner in which the various acts of parliament have been framed. The Standard is often alluded to, but without being defined, or even identified: and it was maintained that any yard measure, of whatever material it might be constructed, having been duly stamped at the Exchequer (in which process, no great pains were taken or required) became immediately a legal standard, however much it might differ from others of

> a similar designation and bearing the same mark of authenticity.[2]

Such critical appraisals were numerous before the creation of the imperial system of weights and measures in the early nineteenth century. Faulty workmanship, inadequate storage facilities, superficially conceived or improperly executed legislation, and haphazard enforcement procedures were cited habitually as the principal causes for the metrological impass. The inability of revisionist policy to correct the outstanding problems was apparent to many government and scientific leaders by the later eighteenth century. Revisionism failed. A revolution was on the horizon.

New Technology

The second half of the eighteenth century was one of intensive technological innovation, and hundreds of new devices, instruments, and scientific procedures appeared. Jesse Ramsden (1735-1800) was perhaps the leading inventor and instrument maker of this era. His dividing engine introduced about 1777 enabled the circles of astronomical and surveying instruments to be divided to an accuracy previously unobtainable. The two large theodolites made for the principal triangulations of Great Britain and Ireland are considered to be his finest achievements. Several sights were taken with them during the surveys at distances exceeding one hundred miles, an impossible distance previously. His other contributions of metrological importance were in his refinements to micrometer microscopes and sextants. Equally important was the discovery of the achromatic object glass by Chester Moor Hall (1703-71) and its development by John Dolland

(1706-61). This was a giant step in the perfection of optical instruments since the color-corrected lenses permitted shorter, more easily handled telescopes.

Two inventors of international stature who would participate directly on metrological committees were John Bird (1709-76) and Jonathan Sisson (1694-1749). Bird specialized in the manufacture of large quadrants and telescopic instruments, and his best work made its way to the observatories such as his eight-foot quadrant constructed for Greenwich. Sisson won fame for an eight-foot quadrant made for the private observatory of George III (r. 1760-1820) at Kew. William Herschel´s (1738-1822) contributions to the development of the reflecting telescope were also crucial. His discoveries created a great demand for astronomical instruments, some of which he made himself, others were made by his associates.

Finally, many of the precision instruments used by metrologists in the later eighteenth and early nineteenth centuries were constructed by a firm founded by Edward Troughton (1754-1835), a famous horologist and metrologist. His transit instruments and equatorials were employed all over Europe and were of great importance in the scientific experiments preceding the establishment of the imperial and metric systems.

Hanoverian Standards

As seen above, substantial progress had occurred in many scientific and engineering fields by the middle of the eighteenth century. As theory and technology expanded, the need for more precise and sophisticated weighing and measuring apparatus, especially as regards primary reference standards, grew more acute. In 1742 the Royal Society, recognizing this urgency, commissioned George Graham (1675-1751), an eminent horologist and scientific instrument maker, and Sisson, the London instrument designer, to make a more accurately subdivided linear measure than the one produced in the revisionist era. This team of Graham and Sisson constructed two finely engraved flat brass bars, 42 inches long, 1/2 inch wide, and 1/4 inch thick. On each of them the lengths of the existing tower standard yard fashioned in 1720 and Elizabeth's standard yard of 1588 were carefully marked off by delicately scratched lines. To make these markings, the team was provided with beam compasses: one of them had parallel cheeks for taking the lengths of standard rods, a second with rounded ends for taking the length of the hollow beds, a third with fine points fitted with an index showing the 800th part of an inch. The two bars were sent to the Academy at Paris where Charles Dufay (1698-1739) and Jean Antoine Nollet (1700-70) marked on them the French half-toise of 3 pieds de roi (38.367 in). One bar was retained in Paris and the other returned to the Royal Society. These bars were exchanged between England and France so that measurements made for scientific purposes in one country could be expressed accurately in terms of the weights and measures of the other. The Royal Society rod is now at the London Science Museum and, in addition to the English and French measurement markings, is inscribed with the words "Royal Society 41" (see Figure 8).

Sixteen years after the issuance of this linear measure, a committee of the House of Commons was appointed to conduct a general examination of all the standards kept at the government's Exchequer depository and those at the Guildhall, Founders Hall, the Watchmakers Company, and the Tower of London. Under the chairmanship of Lord Carysfort, and assisted by Lord Macclesfield, the president of the Royal Society, and by several astronomers and mathematicians, the report of this committee of 1758 contains some vital information respecting eighteenth-century weights and measures.[3] The panel discovered that there were considerable variations among standards of the same denomination (for example, one avoirdupois pound compared with any other avoirdupois pound) and among the aggregates (for example, one avoirdupois 2 pound standard weighed against two avoirdupois 1 pound weights). Product variations received special condemnation from the committee members, especially when they found large discrepancies among standards of the same denomination. For instance, the customary ale and beer gallon was 282 cubic inches, while the legal wine gallon of the Exchequer was 231, and the only existing wine gallon kept at the Guildhall, a mere 224. Although the committee pleaded for the adoption of this smaller wine gallon because it was probably more favorable to taxation, their appeal had little chance of success since the gallon of 231 cubic inches had been legalized previously as the standard in 1707. Nothing was done about equalizing the various liquid measures at one common cubic capacity.

The Carysfort Committee, however, did have two noteworthy successes. Under its direction new standards for the troy pound and yard were constructed in 1758 and 1760 respectively. Legalized in 1824 as the primary reference standards for length and weight, they would be destroyed by the Great Fire that ravaged the Houses of Parliament in 1834. The yard standard, designed and executed by Bird, was a solid bronze bar, 1.05 inches square, and 39.73 inches long. Near each end of the upper surface, and in the middle of a cylindrical hole sunk to the mid-depth of the bar, were inserted gold pins or studs. Each pin was marked with a point or dot to indicate that the distance between them was exactly 36 inches. This standard was gauged against the Exchequer yard-bar of 1588 and corrected to more precise dimensions by the Royal Society's scientific standard of 1742. One of the copies of this yard was marked "Standard Yard, 1758" and presented by the Committee to the House; another was made for common use and sent to the Exchequer. The Bird standard was gauged, in case of loss, to the length of a pendulum beating seconds at the latitude of London in a vacuum. The troy pound was part of a set of hollow and flat brass weights. The hollow pile in denominations of 1/2 to 12 ounces consists of cup weights that fit snugly into each other. All of the weights in the flat pile in denominations of 1 grain to 5 pennyweights have square surfaces with the exception of one octagonal grain weight.[4]

A second report of the Carysfort Committee appeared in 1759. Consisting principally of proposals for legislative regulations, it aimed at facilitating the equalization of weights and measures by establishing proper methods of checking and authorizing the legal standards. In 1765, two bills based upon the committee's deliberations were brought before the House of Commons by Carysfort. The first bill recommended that based upon

Bird´s yard-bar, the ell should contain 1 1/4 yards; the perch, 5 1/2 yards; the furlong, 220 yards; and the mile, 1760 yards. Using the same physical standard, the superficial perch should contain 30 1/4 square yards; the rood, 1210 square yards; and the acre, 4840 square yards. In capacity measurement the quart should be 70 1/2 cubic inches; the pint, 35 1/4 cubic inches; the peck, 564 cubic inches; and the bushel, 2256 cubic inches. In the second bill all standards were to be deposited in the Court of the Receipt of the Exchequer and protected by the Chancellor and the Chief Baron. If anyone wished to use them for correcting and adjusting other weights and measures, he would apply to the Chief Baron and if permission were granted all operations would be conducted under proper Exchequer supervision. (For examples of weight standards constructed both before and after these recommendations, see Figures 9 and 10.) In addition to reducing tne corpus of weights and measures law to one basic enactment which should have been done during the revisionist era, the committee advised that a full-time and carefully staffed inspectorate should be organized and stationed permanently in London.

These two bills were read twice and printed on May 8, 1765, but Parliament adjourned for a temporary recess on May 25 and the issue was dropped. Parliament took no immediate action on these bills during the next session, but many of their provisions would become law after 1824.

The last weights and measures committee of the eighteenth century was appointed in 1790. No minutes of its proceedings are recorded.

The failure of Commons to act on Carysfort´s petitions of 1765 and the probable inability of the committee of 1790 to bring to a successful conclusion any of the tasks entrusted to it were not to be portents for future events. The years between 1790 and 1824 were the most active in the areas of scientific experimentation, government sponsored programs, and committee assignments in British metrological history. Not only the most active, but the most successful, for it was during this period that the imperial weights and measures system was created and legalized.

During the last decade of the eighteenth century and the first fourteen years of the nineteenth, numerous scientists and scientific societies began to provide themselves with weights and measures standards. Sir George August Shuckburgh-Evelyn (1751-1804), for instance, had Troughton construct for him a yard standard that he compared later with the government´s standards. This "Shuckburgh scale" would figure prominently in all future standards experiments by the government. Troughton made a similar measure for himself and introduced the newly perfected micrometer microscope into metrological sizing operations. The latter yard-bar was built in conformity with one that Bird had made for the Assay Master of the Mint. Still another was constructed for General William Roy (1726-90), and it was used by this member of the Survey Ordnance Office in many official examinations. Other examples could be documented.

Pre-Imperial Reform Proposals

While these technical successes continued to mount, researchers were achieving pioneering breakthroughs in other branches of the physical sciences. As soon as the measurements of the earth began to be computed

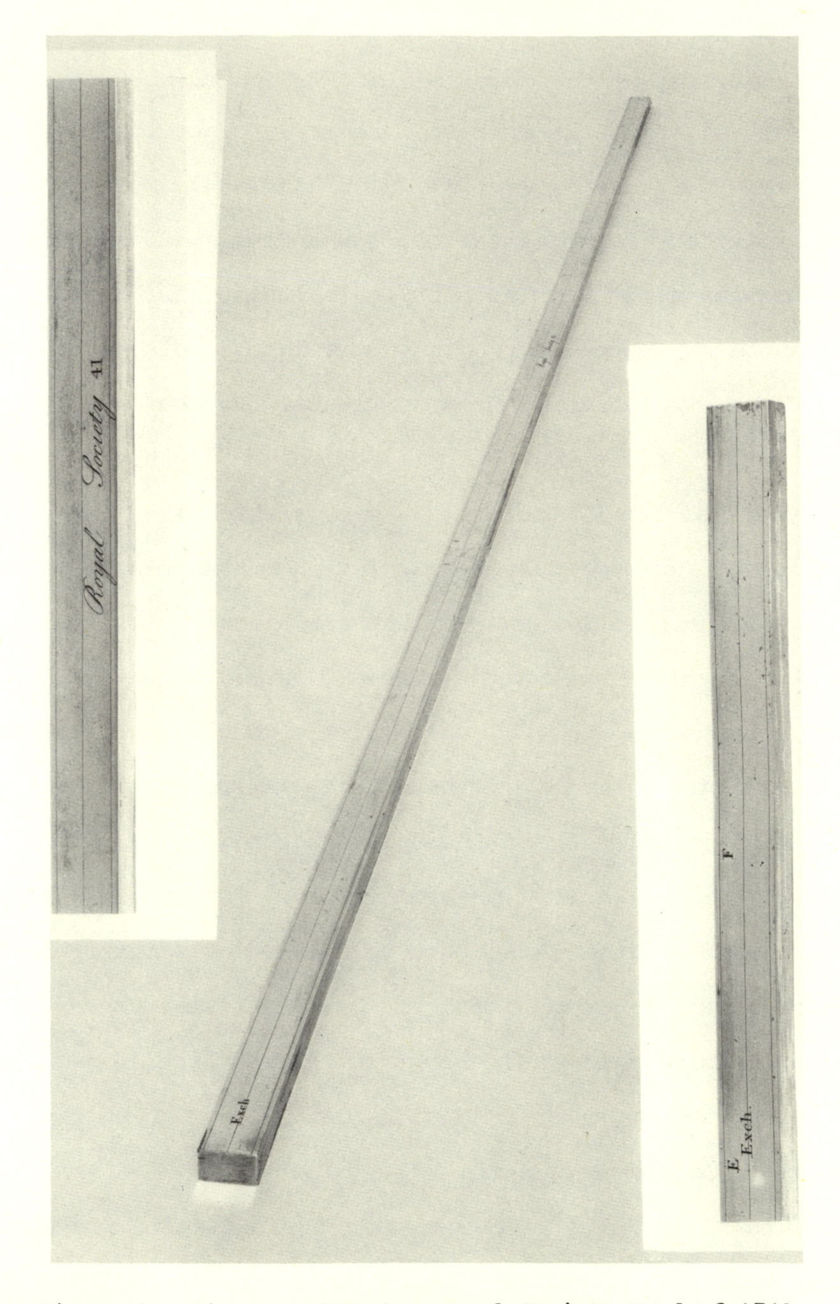

Figure 8. Sisson and Graham Royal Society Yard of 1742.
(lent to Science Museum, London by the Royal Society)

Figure 9. Avoirdupois Cup-Shaped Weights of George I.
(Crown Copyright. Science Museum, London)

Figure 10. Avoirdupois Cup Weight (1773) of George III: 28 pounds, 14 pounds, 7 pounds, 4 pounds, 2 pounds, 1 pound, 8 ounces, 4 ounces, 2 ounces, 8 drams, 2 drams.

(Photo. Science Museum, London. By courtesy of University museum of Archaeology and Ethnology, Cambridge) with some proficiency, the French proposed a standard measure based on the ten-millionth part of a quarter of the meridian. The Parisian savants expected such accurate precision that future measurements would not produce even a microscopic alteration in the length of their new meter. In England the pendulum was being given careful consideration. The early ideas of Wren and Huygens suggesting that the pendulum be adopted as the standard of linear measurement resurfaced. Since the second of time is determined by the motion of the earth, it was conceived that the length of the seconds pendulum in a given latitude would be an invariable quantity that could always be recovered or duplicated.

The most well-known exponent of the seconds pendulum was Sir John Riggs Miller (fl. 1760-1800). A baronet, barrister, and professional soldier, he was intensely interested in mercantile matters and while a member of Parliament from 1784 to 1790, he worked out what was considered then a bold and novel approach to weights and measures reform; one which suggested that the pendulum was the most likely device by which one could establish a permanent standard. But during these critically eventful years when the French were undergoing violent political unheavals and, at the same time, supplanting their old and immensely complex metrology with the painstakingly researched and elaborately tested metric system, there was expectation that France and England would cooperate in a joint venture to give this new system of weights and measures an international character. Letters were sent back and forth between the Royal Society and the Academy for the hoped-for entente. When the cooperation fizzled out, the Miller plan was placed before the House of Commons on April 13, 1790.

At the outset, however, it must be pointed out that some of Miller´s fundamental ideas and much of his data were borrowed from others. Besides owing a great debt to the compilers of the *Report of the Committee of the House of Commons in 1758*, in the monographic account of his plan entitled *Speeches in the House of Commons Upon the Equalization of Weights and Measures of Great Britain* published in London in 1790, he credits the French Bishop of Autun (the celebrated Talleyrand) for several metric conclusions arrived at through an exchange of letters; George Skene Keith (1752-1823), Minister of Keith Hall, Aberdeenshire, for correspondence relating to a plan that Keith had been considering proposing himself; and Sir James Steward who had supplied him with an unsigned manuscript (dated 1760)—*A Plan for Introducing an Uniformity of Weights and Measures Within the Limits of the British Empire*—found by Steward among his late father´s personal papers. Also, in setting up the groundwork for his study, Miller in 1789 had addressed a general circular letter to the chief magistrates of all cities, corporations, and borough towns of Great Britain soliciting their assistance. Most of them responded and he added substantially to his knowledge of local metrological practices. Finally, in 1790 he requested all English and Welsh clerks of the market to draw up and submit to their sheriffs a comprehensive listing of the weights and measures currently employed within their territorial jurisdictions. After the compilations were finished they were sent to the House of Commons for examination. Scotland was excluded purposely since their local units were not recognized after the passage of the Act of Union.

There were three motives behind Miller's actions: to demonstrate to Commons the enormous complexity of local units; to analyze and systematize the principal reasons for their continued use; and, most important of all, to propose what could be done to eradicate them and bring about standardization. After discussing briefly the first two points, he concentrated on point number three.[5] Before actually establishing his views on the latter, however, he became the first English metrological reformer to recommend that the existing values and proportions among weights be altered completely to conform to the following decimal scale:

Decimal Multiple	Denomination	Troy Grains
10 grains	1 scruple	10
10 scruples	1 dram	100
10 drams	1 ounce	1000
10 ounces	1 pound	10,000
10 pounds	1 stone	100,000
10 stone	1 hundred	1,000,000
10 hundred	1 ton	10,000,000

It is unfortunate that he did not extend this decimalization to linear and capacity measures, since he would have constructed a complete decimalized metrology almost a decade before the French. He concluded by calling for the elimination of all other weights and suggested that grain products should be sold by weight instead of by capacity measure. His lone recommendation concerning linear measurement was that steel yards be replaced as the physical standards. He never said what their replacements should be.

Eighteenth-century metrologists were immensely impressed with the profound discoveries of the Age of Science, and as representatives of the Enlightenment, they wished to incorporate any new data or theories into their weights and measures reforms. Riggs Miller was no exception. In searching for a more precise and a more empirically verifiable basis for the establishment of physical standards, he concluded that there were four possible approaches and that the last of these, in his estimation, was clearly the best. First, it would be possible to adopt a standard based upon the mass of a drop of liquid (ideally distilled water) weighed at a certain controlled temperature. Presupposing that every drop were of equal mass, a given number of drops might be selected for a standard ton weight, and the side of the vessel that contained this amount might become the standard for linear measure. To insure that the drops were exactly equal, the distance and velocity of their fall would have to be controlled. Two people were necessary: one to keep the temperature uniform; another to count the time consumed in the fall of each drop. If these conditions could be realized, the particular advantage of this method would be that since the standard was taken from the cubic to the linear measure, the error, if any, would be decreased in making measures of length, while any error in a standard raised from a linear measure is greatly increased for capacity or volume measures. Its major drawback was that a drop is an extremely small mass. If any error were made in its dimensions, that error would be compounded tremendously in quantities as large as a ton.

The second approach—measuring the distance covered by a falling body in one second of time—was the least practical. The principal argument against it was that although it was an excellent idea in theory, there was no experimental method known that could be used to determine the distance with absolute precision.

If the first approach were faulted for its dependence on too small a mass, a standard taken from a degree of a great circle of the earth would be far too large to ascertain with accuracy, and it would have to be subdivided too many times to derive workable quantities, Again, even if such a standard were achieved it would not be universal since a degree measured in another latitude would be either longer or shorter, and even if measured within the same latitude, the gravitational effects of different terrains would produce variances in the lengths of any two degrees selected for measurement.

The problems inherent in these three approaches, in Miller´s mind, forced him to become one of England´s most vociferous advocates of a standard based upon the length of a pendulum vibrating seconds in the latitude of London. This standard had three outstanding attributes: structural simplicity, relative ease of execution, and scientifically recognized accuracy. Not actually conducting any experiments himself, he came to regard the conclusions reached by four scientists—operating independently of one another—as the point from which to begin selecting his ideal measure of length.

Table 3.1

Seconds Pendulum Lengths

Scientist	Description	Pendulum Length in Inches
Emerson[6]	measured at different latitudes	39.131
Desagulier[7]	measured in London	39.128
Graham	1722 experiment made with a standard English foot	39.126
Whitehurst[8]	1. deduced from interval between two pendulums	39.1196
	2. corrected by others who made a reduction for the small rod of his pendulum	39.1187

Since Graham's was approximately the median among the other lengths, Miller considered it the best choice for the moment until a nationally supervised experiment could be held sometime after 1790. Basing his proposal in the House of Commons on this fact, he asked that the London pendulum of 39.126 inches be accepted for the time being as the standard for linear measure; the square of this length to represent the standard for area or superficial measurement; its cube the standard for capacity or volume measurement; and the latter filled with rain water of a certain temperature as the standard ton, from which all other weights were to be derived.

Miller´s proposal was read before Commons in 1790 and for much of that year was debated fiercely. It was not enacted into law but its influence on later imperial commissions was considerable as will be seen.

Four years after the publication of the Miller monograph, a second metrological reform program was inaugurated by William Martin (fl. 1760-1800). Aside from serving as the treasurer to the Aire and Calder Navigation of Wakefield, nothing whatever is known concerning his background or the circumstances surrounding the issuance of a book written by him and published in London in 1794. Entitled An Attempt to Establish Throughout His Majesty´s Dominions an Universal Weight and Measure, Dependant on Each Other, and Capable of Being Applied to Every Necessary Purpose Whatever, it concentrated exclusively on altering the contemporary systems of avoirdupois weight and of dry and liquid capacity measure. Perhaps realizing the futility of proposing a radical alteration in the standard for the existing system of weights and measures, he eschewed Miller´s pendulum emphasis and opted for continuance of the contemporary method of establishing prototype standards for all weights and measures.

Beginning with an avoirdupois pound totaling 20 ounces of 437.5 grains each (the national standard was 16 ounces of 437.5 grains each or 7000 grains in all), Martin restructured thoroughly all of the values and proportions in the avoirdupois system to create a semi-decimal system of weights.

Table 3.2

Martin Weight System

Multiple	Denomination	Troy Grains
20 grains (imaginary)	1 pennyweight	21.875
20 pennyweights	1 ounce	437.5
20 ounces	1 pound	8750
100 pounds	1 hundredweight	875,000
20 hundredweight	1 ton	17,500,000

The new avoirdupois pound was 25 percent larger than the former standard, while the hundredweight and ton were slightly more than 11.6 percent larger.

In order to establish one common capacity measure, Martin recommended that the bushel should contain one cubic foot or 1728 cubic inches (the national standard was 2150.4 cubic inches). The gallon, one-eighth of this measure, would be 216 cubic inches; the quart, 54; and the pint, 27. The gallon would be considered the primary standard (called an integer by Martin) from which all other standards were to be derived. For the convenience and accommodation of merchants, he suggested that there be three separate systems of capacity measurement. All of them, however, were based on the standard gallon.

Table 3.3

Martin Capacity Measures

Multiple	Denomination	Cubic Inches
	Dry or Grain Measure	
2 pints	1 quart	54
2 quarts	1 quartern	108
2 quarterns	1 gallon	216
2 gallons	1 peck	432
4 pecks	1 bushel	1728
10 bushels	1 quarter	17,280
	Wine Measure	
2 pints	1 quart	54
4 quarts	1 gallon	216
67 1/2 gallons	1 hogshead	14,580
2 hogsheads	1 pipe	29,160
2 pipes	1 tun	58,320
	Ale and Beer Measure	
no submultiple	1 gallon	216
24 gallons	1 kilderkin	5184
2 kilderkins	1 barrel	10,368
1 1/2 barrels	1 hogshead	15,552

Table 3.4 records the proportions that existed between Martin´s revised units and those in use prior to the imperial system. The third column records the percentile of deviation from the contemporary standard.

Table 3.4

Differences Between the Martin Proposal and the Contemporary System of Weights and Measures

Unit	Martin´s System	
bushel	1728 cu in	0.8035714 of 2150.4 cu in
quarter	17,280 cu in	1.0044642 of 17,203.2 cu in
wine gallon	216 cu in	0.9350649 of 231 cu in
wine tun	58,320 cu in	1.0018552 of 58,212 cu in
ale and beer gallon	216 cu in	0.7659574 of 282 cu in
ale and beer barrel	10,368 cu in	1.0212766 of 10,152 cu in

Martin´s plan failed to win much support. At least the Miller proposal had been given a parliamentary hearing and had been considered for adoption by the Lower House. Its ultimate demise was due not to any blatant weaknesses, but to its dependence upon the seconds pendulum as the basis for Britain´s physical standards. This was just too radical a departure from traditional practice. Martin, however, seems to have been largely ignored by Commons since his plan never was debated in any of their sessions. Of course, it could be argued that Miller had an obvious advantage in this respect since he was a member of Parliament at the time that he made his proposal. But Martin apparently was never able to bring his ideas to the attention of the general public. He is never mentioned in any kind of contemporary source and his work is never quoted or referred to in the speeches, pamphlets, articles, and monographs of later metrologists. Perhaps his plan suffered from the fact that it was not radical enough. He advanced no complete decimal scheme at a time when decimalization was in vogue. He did not extend his reform plan to linear and superficial measurement. He offered no alternative to the existing mode of determining physical standards. He compiled no supporting evidence for his reforms from government records, from other metrological specialists, or from other scientists. At best, his remedies were vague and incomplete. If Miller had gone too far for late eighteenth-century England, Martin did not go far enough.

A third reformer of far more importance than Martin was Miller's brilliant contemporary and friend, George Skene Keith. Born near Aberdeen in 1752, he was the eldest son of James Keith and a descendant of Alexander Keith, third son of the second Earl Marischal. He graduated from Marischal College and the University of Aberdeen in 1770, was licensed by the presbytery of Aberdeen in 1774, and by 1803 had received the Doctor of Divinity degree.

Besides his many other interests, for over thirty years Keith investigated various methods for reforming weights and measures and supported strongly, just as Miller had, the adoption of the seconds pendulum as the basis for all British physical standards. His earliest proposal was laid before a committee of Parliament by Miller in 1790 but it, like Miller´s, failed to win immediate acceptance. Sir Joseph Banks (1743-1820), a distinguished member of several imperial weights and measures commissions and a fellow of the Royal Society, spoke enthusiastically about Keith´s first pamphlet defending the pendulum proposal, published in London in 1791, and entitled a Synopsis of a System of Equalization of Weights and Measures of Great Britain.

Keith continued to defend his proposal in the years that followed, and in 1817 he drew upon more than a quarter-century of intensive research to publish in London his most important work—Different Methods of Establishing an Uniformity of Weights and Measures. Looking for remedies to prevent further multiplication of state and local units, Keith, like so many metrologists before him, saw the solution in the establishment of one general standard. Ideally, this standard should be a linear measure, connected with some natural constant that is permanent and invariable. From this standard all other measures would be derived: its square, just as in Miller´s proposal, becoming the standard for superficial measures, and it cube the standard for capacity measures. Further, since Keith believed that "pure water" at a controlled temperature was always of the same specific gravity, the cube of the linear standard, filled with distilled water, ought to be the standard for weight. If this were unacceptable a cylinder could be substituted, one whose diameter and weight corresponded with the linear measure. The weight of water in a hollow cylinder could represent the standard, or one could base it upon the amount of water displaced when a solid cylinder was immersed in some other container.

There were only two acceptable ways, argued Keith, to establish a standard. The first, and the most scientifically reliable, was to take a standard from nature that, at any future date, could be verified easily and accurately. The second was to adopt any of the present standards and fix its relation to a natural standard. If one followed the first method, all weights and measures would be aliquot parts or exact multiples. In the second they would be aliquant parts or exact submultiples. In either case, it was imperative that one connect the standard, either directly or indirectly, to some natural constant, otherwise it could not be preserved unaltered for any length of time. The first method, obviously, would be the most permanent, the most accurate, and the one most likely to be adopted by other nations, while the second would cause fewer transitional problems for the English and would be more readily understood.

Concentrating almost exclusively upon method number one, and using data originally collected by an imperial weights and measures commissioner, John Playfair (1748-1819) of Edinburgh, a philosopher and mathematician, Keith recognized three possible standards: the length of the seconds pendulum in a given latitude; a length derived from a degree of the meridian in a given latitude; and a length based upon the distance required to produce a certain, predetermined fall in a barometric reading. The last two were considered inferior. The barometer registered different results in different levels of the atmosphere and seldom could be ascertained with sufficient accuracy within 1/600 part; often not within 1/300 part. An error of 1/300 of its length produced an error of one percent on a solid body or on a standard weight or capacity measure derived from it. The length of a degree of latitude entailed considerable expense and many verification problems. Besides, either in the original mensuration or in the subsequent verification, the precise length of such a degree could never be determined with the accuracy requisite for an invariable standard since the effects of mountains, oceans, and other topographical influences were never constant. Even if such problems could be resolved, the standard derived would not be the same in both hemispheres, or even in the same hemisphere if measured in different degrees of longitude.

Rejecting both of these approaches for the reasons cited, Keith, as his colleague Miller had done, opted for the pendulum. Basing his argument on data once again obtained from Playfair, he stated that the pendulum vibrating seconds at London measured, in two readings, 39.126 and 39.130 inches. The difference between them was 0.004 inches or about 1/10,000 part on a linear measure; more exactly, 1/9782 part on a linear standard and 1/3261 on a weight or capacity measure derived from it. If the true length were found ultimately to be 39.128 inches, the difference would only be one-half of these insignificant fractions.

To counteract arguments raised against this type of standard, Keith noted that differences in the length of a seconds pendulum measured in different latitudes were extremely slight. For example, its length at

London was approximately 1/1800 part shorter than at St. Petersburg, while it was 1/1500 part longer than at Constantinople. Secondly, there was no truth in the charge that it was difficult to determine the precise length of the pendulum. In the cylindrical or uniform model, it was simply two-thirds of the length of the rod. In other models there were additional steps required but they never necessitated elaborate calculations.

After dismissing the French metric system from consideration because, among other things, it was based upon the ten-millionth part of the quadrant of the meridian (which he, unfortunately, considered too difficult a measurement), Keith contemplated what alternative plan could be substituted if Parliament continued to oppose the adoption of the pendulum. Only one possibility seemed plausible. A standard could be established by selecting one of the most common weights and measures and building a new system around it. It was fortunate that there was a direct correlation between the cubic foot and 1000 avoirdupois ounces. A cubic foot of distilled water at a temperature of 56 1/2° F. contained exactly 62 1/2 pounds or 1000 avoirdupois ounces. Also, an English quarter contained almost 10 cubic feet; the quarter raised from the standard bushel of Excise was only 1/225 less; and that raised from the standard gallon of the Exchequer was 1/270 more. It is the correlation of 1000 avoirdupois ounces with the cubic foot, and the approximate correlation of 10 feet with the quarter, that offered hope for this alternative plan.

Regardless of these correlations, however, the pendulum method was far superior. A pendulum 12 inches long vibrates almost exactly 156,000 times a day, or 6500 times an hour, or 108 1/3 times a minute. Consequently, a pendulum 4 feet long would vibrate 78,000 times a day, 3250 times an hour, or 54 1/6 times a minute. By constructing a machine and proportioning the rods and balls of two pendulums—one of 4 feet, the other of 1 foot—the difference between the lengths of these rods could be made an English yard, not only in London but in any part of Europe. Thus an invariable linear measure could be obtained. The yard's length could also be derived from a pendulum that vibrates 90,060 times in a solar day. The former procedure, however, was much simpler.

Using the yard obtained as the only primary standard, Keith suggested that the following changes be made among capacity measures and weights.

Table 3.5

Keith Unit Revisions

Denomination	Description	Ounces of Distilled Water at 56 1/2^{O} F.
CAPACITY		
tun	40 cubic feet or 69,120 cubic inches	40,000
quarter	1/4 tun or 10 cubic feet or 17,280 cubic inches	10,000
firlot	1/5 quarter or 2 cubic feet or 3456 cubic inches	2000
half-firlot or	1 cubic foot or 1728 cubic	1000

Table 3.5 (continued)

Denomination	Description	Ounces of Distilled Water at 56 1/2° F.
cube	inches	
gallon	1/10 firlot or 1/5 cube or 345.6 cubic inches	200
half-gallon	172.8 cubic inches	100
pint	1/10 gallon or 34.56 cubic inches	20
half-pint	17.28 cubic inches	10
glass	1/10 pint or 3.456 cubic inches	2
half-glass	1.728 cubic inches	1
WEIGHTS		
pound	20 ounces	NA
hundredweight	100 pounds	NA

Keith advocated retaining the Winchester bushel for dry products since it was so popular and since the malt tax was based upon it. He borrowed the firlot from Scottish metrology and made it nearly identical to the Linlithgow standard of 3230.2 cubic inches. The latter was a serious mistake since Scottish weights and measures were illegal after 1707. The new gallon was almost twenty-five percent larger than the ale gallon of

Excise and almost fifty percent larger than the wine gallon. The pint was nearly equal to the ale pint. His only suggestions concerning weights were that there be no irregular subdivisions among them, and that they be proportioned decimally. He never defined any weights except the pound and hundredweight, nor did he specify whether any were to be eliminated. His plan lacked completeness and, as such, suffered the same fate as Martin's.

The most comprehensive pre-imperial reform was introduced in the second decade of the nineteenth century by Francis Perceval Eliot, an English financial expert and economist. Born about 1756, he entered the civil service in the late 1770s, and was for seventeen years, from 1806 to the time of his death in 1818, one of the commissioners of audit at Somerset House. He was also intensely interested in the volunteer yeomanry service, was successively major and colonel of the Staffordshire volunteer cavalry, and wrote considerably on the subject. Besides these pursuits, he devoted more and more of his time in later life to the study of historical metrology. In 1814 in a collection of Letters on the Political and Financial Situation of the Country...Addressed to the Earl of Liverpool published in London, Eliot proposed a program whose central theme was the establishment of "one uniform weight and one measure."[9] He believed that this long sought after goal could finally be realized if his recommendations were enacted into law.

The early sections of his Letters—published originally in a series of short articles in the journal Pamphleteer—concentrate on the elimination of all apothecary and troy weights. Eliot saw no need for the apothecary system since druggists customarily bought and sold their merchandise by avoirdupois weight. Furthermore, the apothecary pound consisted of the same number of grains as the troy, only its intermediate divisions were different. It would have been far wiser, he argued, if druggists had employed the troy ounce and had subdivided it into halves and quarters of 10 and 5 pennyweights each, the latter again halved and quartered into 12 and 6 grains. As it stood, there were three independent systems: one for compounding weight units of drugs; another for the Mint; a third for everything else. Since the avoirdupois system was overwhelmingly the most utilitarian, it alone should be employed. It could be restructured to accommodate British pharmaceutical and minting requirements, and since its pound was already divided into ounces and drams, all that had to be done was to add several denominations smaller than a dram and several larger than a pound. In this restructuring, however, it was imperative to maintain throughout the entire system that "...perfect series of squares and cubes, which can alone secure an infinite regularity."[10]

Before carrying this discussion to its logical conclusion, Eliot interjected several recommendations for improving liquid and dry capacity measurement. In the former there were three separate and unrelated systems—ale, beer, and wine—with no consistency in their intermediate divisions. To resolve this dilemma, he proposed that one system be applied to all liquids and that it be based upon the Winchester wine pint of 28 7/8 cubic inches, the latter containing exactly 1 pound of pure water. Likewise, in dry measure, the pint should contain the identical cubic capacity and be filled in such a way that its contents weigh exactly 1 pound. Each pint would be striked or heaped depending on what type of grain it contained. (Like Keith´s acceptance of the firlot, an illegal Scottish unit after the Act of Union, this was another serious mistake since such measures invited circumvention of the standards and had been traditionally one of the major bones of contention in commercial relations.) Comparatively, the new bushel—containing 64 of these pints—would be 1848 cubic inches and contain 64 pounds of grain.

Returning to the problem of how to restructure the avoirdupois system, Eliot suggested that the first step must be to increase the weight of the pound—prophetically renamed the "imperial"—to 8192 grains, each of the latter equal to "...one grain of the largest and heaviest wheat.[11] (The most obvious problem here, of course, was what constituted the largest and heaviest wheat grain; this was still another leftover from medieval legal parlance that caused so many horrendous obstacles in establishing standards in past centuries.) Drawing almost exclusively upon experiments conducted by others, he calculated that one of these new grains equalled 0.8545947265625 troy grain or, conversely, that 1 troy grain equaled approximately 1 1/3 new grains. Building upon this basic standard, he recommended that 32 "imperial" grains make the dram; 512, the ounce; 8192, the pound; 131,072, the stone; 1,048,576, the hundredweight; and 16,777,216, the ton. The proportions between each of these units are outlined in Table 3.6.

Table 3.6

Proportions Among the Weights in the Eliot Proposal

drams	ounces	pounds	stone	centals	tons
1					
16	1				
256	16	1			
4096	256	16	1		
32,768	2048	128	8	1	
524,288	32,768	2048	128	16	1

Because this scale was dependent upon multiples of 4, it was well suited for simple arithmetical reckoning. The higher weight assigned the stone was in keeping with this general pattern. The new ton of 2048 pounds was much nearer the original weight of 2000 pounds than the contemporary one of 2240 pounds, while, at the same time, it avoided all of the irregular divisions so common in the older systems. Lastly, the grain could be squared or cubed in multiples of 4 to satisfy the needs of druggists, jewellers, and moneyers. Eliot, regrettably, never suggested any names for these smaller units.

In presenting his complete list of "imperial" dry and liquid capacity measures, Eliot laid special emphasis on the fact that if 8 avoirdupois pounds of pure water should ever be found to occupy a greater or lesser space than 231 cubic inches, the solid contents of the gallon could be altered accordingly without interfering at all with the proportions among the various measures. It would only be necessary to adjust the cubic capacities of these other measures so that they corresponded in direct proportion. Besides abolishing the wey and the last, he united the ton for dry measure with its counterpart among the "imperial" weights, thus ending the age-old confusion between these two units. Tables 3.7 and 3.8 show the proportions among Eliot´s dry and liquid capacity measures.

Table 3.7

Proportions Among the Dry Capacity Measures in the Eliot Proposal

pints	quarts	gallons	pecks	bushels	sacks	quarters	tons
1							
2	1						
8	4	1					
16	8	2	1				
64	32	8	4	1			
256	128	32	16	4	1		
512	256	64	32	8	2	1	
2048	1024	256	128	32	8	4	1

Table 3.8

Proportions Among the Liquid Capacity Measures in the Eliot Proposal

pints	quarts	gallons	firkins	kilderkins	bbls	hogsheads	pipes	tons
1								
2	1							
8	4	1						
64	32	8	1					
128	64	16	2	1				
256	128	32	4	2	1			
512	256	64	8	4	2	1		
1024	512	128	16	8	4	2	1	
2048	1024	256	32	16	8	4	2	1

Eliot´s plan aimed essentially at altering the existing system of weights and measures. Devoid of decimalization and other radical changes, it was an attempt to get the Parliamentary commissioners to reform the contemporary system by removing superfluous weights and measures and by adjusting the remaining ones to achieve internal consistency and conformity. The latter aspect was its most novel feature. Unfortunately, he, like some of his contemporaries, left too many final determinations of the actual standards upon which reform was based, and the actual scientific experimentation, to other unnamed individuals. How was Parliament to conclude that the foundational work would ever be completed, or that the final results would corroborate the initial findings of the seminal plan? But, to Eliot's credit, simplicity was the keynote of his endeavors. His use of the name "imperial" was indeed a prophetic move for it was the same one (perhaps by deliberate parliamentary design) used shortly thereafter by London for a government-sponsored system that became part of British law in 1824. It was unfortunate that only in the latter respect did Eliot seem to have had any appreciable influence.

A number of conclusions can be drawn from this examination. Besides the many faults mentioned in preceding discussions, all of the plans advanced before 1824 suffered from the fact that they were undertaken as independent actions by private citizens and were not sponsored or sanctioned directly by the central government. Since they responded vociferously to centuries of governmental recalcitrance in effecting significant and wide-spread reform, London viewed them as vicious attacks and generally spurned them. All of the major changes in British metrology between the Tudor and the Hanoverian periods were brought about through the work of government-sponsored commissions, and the dominant characteristic of these panels was conservatism. The reformers generally were not conservative (in the scientific sense), and they had attacked the accomplishments of the commissions as frequently as they did the central government. Hostility and competition among these groups, and the fact that the reformers could not propose a united front and a coordinated program, worked almost solely to the latter's detriment.

Further, any plan that utilized decimal scales in whole or in part was viewed by London as a French or metric scheme, and the government was well aware of the difficulties that the revolutionary governments of France were having in enforcing compliance to their new system. In fact, there was so much disfavor shown the metric system in France after its promulgation in the last decade of the eighteenth century that Napoleon Bonaparte (1769-1821) was forced, in 1812, to permit use of the weights and measures of the ancient regime and to discontinue temporarily Paris' efforts of achieving nationwide uniformity. It was not until 1840 that the metric system was again made mandatory in France (see Chapter 5). Such difficulties made London political leaders wary of decimalization and seriously crippled such metrological reform proposals even before they were brought to the attention of the government. Also, the English ministers and parliament could look to the American experience and see that the metric system had fared no better there. On at least four occasions, proposals were made for an American decimal metrology during the last decade of the eighteenth and the first decade of the nineteenth centuries, and on each occasion failure resulted. Even the backing of Thomas Jefferson (1743-1826) was insufficient in bringing a decimal scale to American weights and measures (see Chapter 4). Hence, London could claim that decimalization was unpopular everywhere and therefore would stand little chance of success in the British Isles.

Thirdly, pendulum reliability was mistrusted by too many of the scientists who served on the government's commissions. Since there was no general agreement as to the exact length of a pendulum vibrating seconds in the latitude of London, the commissioners were reluctant to construct a new metrological system on it only to find out later that their original figures were incorrect and that substantial changes would have to be made. If considerable funds would have to be expended to create and implement change in the contemporary system, the government wanted absolute assurance that its monies were well spent, and it was unconvinced that the pendulum method was as scientifically reliable as its advocates claimed.

Finally, there was the powerful influence of custom and tradition. Not only had the contemporary system become part of the national culture, but it was inextricably linked with the phenomenal successes of the Industrial Revolution. England was the ranking industrial nation in the world, and its manufacturing, commercial, and financial leaders argued that too abrupt or radical a change would impair present and future growth (a claim made repeatedly down to the present day). If machine parts had to be changed, or if the dimensions of most of England's exports had to be altered, untold confusion would result in the national economy and the entire nation would be thrown into a recession or a depression. (Similar arguments have been presented by anti-metric spokesmen in the United States throughout the present century.) It was far wiser, they claimed, to proceed cautiously with change and only modify those aspects of the existing metrology that were detrimental to further economic growth. London was convinced by these arguments (notwithstanding the political clout of such powerful lobbies) and by those coming from other sectors of the populace and eventually compromised with the legislation passed in 1824. Moderation rather than revolution was its theme. In the end, the reformers had failed to alter significantly the conservative course of British metrological evolution, even though many of their recommendations would find their way into the new law and those to come later.

Government Reaction

In 1814 the House of Commons appointed a new committee of 23 members to investigate the possibility of incorporating some of the results of this quarter-century of investigation and experimentation into the government's program of finding a more reliable basis for weights and measures standards. The first action of the committee was to recommend that the yard-bar constructed by Bird for the Carysfort Committee be accepted as the official linear standard. This decision was based on information supplied by William Hyde Wollaston (1766-1828) and Playfair who had determined the length of a pendulum vibrating seconds in the latitude of London to be 39.13047 inches, of which Bird's yard contained 36. Both scientists alluded to the necessity of verifying this number, but the members of the committee took it for granted, asserting finally that any expert watchmaker could make a seconds pendulum. (How wrong they were, as later horological evidence would show.) They never stated how that pendulum was to be measured after construction, or at what temperature or pressure it was to be swung.

After completing their work on the linear standard, the committee turned its attention to capacity measurement. Although in theory the original standard of weight is best derived from the measure of capacity (a reformist proposal), the committee felt that, in common practice, it would be more convenient to reverse this order. Hence, they recommended, upon the suggestion of Wollaston, that a gallon containing 10 pounds of pure water should be adopted as a substitute for all of the current standard gallons. One measure would suffice for dry and liquid products. Even though twelve wine gallons of distilled water weighed 100 avoirdupois pounds, Wollaston´s proposal to make a gallon exactly of 10 pounds afforded greater facility in the operation of adjusting the measure since it was not very easy to divide 100 pounds into 12 equal parts with the weights in common use.

Shortly after the committee´s report was made public, doubts arose concerning several of its conclusions. Because of the committee´s failure to take into consideration the important experiments on weights and measures conducted by Shuckburgh, who had published his findings in the Philosophical Transactions as early as 1798, and by John Whitehurst, who had obtained the length of the pendulum by measuring the difference between two lengths vibrating at different frequencies, some particulars in the report were believed erroneous. First, the weight of a cubic foot of water is stated by the committee, taken probably from one of Keith´s works, to be 1000 ounces at 56 1/2° F. Many scientists, including Shuckburgh, claimed that the total weight was less than 1000 ounces even at 39° F., the density at that temperature considered by them to be "maximum." Second, the customary length of the English foot, which had been adjusted almost always at the ordinary temperature of the atmosphere in summer rather than in winter was, in the course of the committee´s experiments, copied from the Genevan scientist Marc Auguste Pictet (1752-1825) and compared with the French standard that was intended to be employed at the freezing point of water. No correction was made by the committee to allow for these differences in technique. Finally, Thomas Young (1773-1829), a highly regarded scientist, had pointed to similar omissions in the earlier reports of government committees. He was particularly annoyed that one of his own computations had been ignored. Long before the committee´s formal schedule began, Young, following the procedures inaugurated by Pictet, and confirmed by the experiments of Bird, Maskelyne, and Landale, the Astronomer Royal, had computed the length of the meter to be 39.3710 English inches instead of 39.3828 English inches, the figure accepted by the committee.[12]

In order, therefore, to resolve these problems, Davies Gilbert (1767-1839), later president of the Royal Society from 1827 to 1830, asked Commons in 1816 to authorize the formation of a second committee that would ascertain the exact length of the pendulum vibrating seconds and that would compare anew the French and English standards of length. Not abiding initially with the full import of Gilbert´s petition, the House directed the Astronomer Royal to perform the necessary operations. When the latter official requested a staff of assistants, Commons ordered the president and council of the Royal Society to nominate a committee that would cooperate in the venture. The members chosen, in addition to the Royal Society´s president and various secretaries, were Gilbert, Wollaston, Young,

Troughton, Banks, Sir Henry Charles Englefield (1752-1822), Henry Browne (fl. 1800-40), Sir Charles Blagden (1748-1820), Captain Henry Kater (1777-1835), Sir John Barrow (1764-1848), and General William Mudge (1762-1820).[13] This distinguished panel began deliberations by discussing which procedures would be followed in conducting the experiments. It was resolved that such matters would be left to the discretion of the individual researchers, and that when they reassembled they would make collateral determinations of the results.

The experiments proceeded cautiously and painstakingly for two years and the results were brought to the attention of the government late in 1818. Before any attempt was made to introduce legislation based on the committee´s recommendations, the ministers of George III thought it advisable to create yet another committee that would correlate the work of the 1816 commission with the overall program of weights and measures reform. This third group, appointed by the Prince Regent early in 1819 by a writ of the privy seal, consisted of Gilbert, Wollaston, Young, Kater, Banks, and Sir George Clerk (1787-1867). Wollaston examined some of the authorized measures of capacity and Young served as committee secretary. Clerk and Gilbert were responsible chiefly for drafting the bills that were submitted to Commons after the formal sessions ended. A law clerk hired by the committee made extracts of the statutes that Young used to draw up an abstract of the present state of weights and measures law. This same clerk also perused the county agricultural reports and compiled, for the first time in English history, an extensive glossary of what he believed were all the weights and measures that had acquired a local popularity throughout the British Isles.[14]

The committee issued three reports, dated June 24, 1819, July 13, 1820, and March 31, 1821. In the opening paragraph of the first report, the commissioners confessed that it would be extremely difficult to effect any radical changes in English weights and measures such as those suggested by Miller and Keith. (As will be seen, the French did not consider a radical alteration to be outside the limits of human perseverance, but, of course, their metrological transformation went hand-in-hand with sweeping political changes.) Consequently, they would proceed with caution in the expectation of altering the contemporary system where necessary rather than replacing it with a totally new system as the French had done during their revolutionary years. In their very first pronouncement, however, they declared that no alteration whatsoever should be made in the standards of length. They did not see any practical advantage in having a linear measure commensurable to some original quantity existing, or which was imagined to exist, in nature, except as it afforded encouragement for common adoption among neighboring nations. The members argued that it was likely that the departure from a long established standard would produce more inconvenience to the internal commerce of the British Isles than it could be expected to save in the operation of foreign trade. They declared further that the subdivisions of weights and measures employed in Great Britain appeared to be far more convenient for practical purposes than the decimal scale, which might perhaps be preferred by some people for making calculations with quantities already determined. Also, the ability to express a third, a fourth, and a sixth of a foot in inches without a fraction was a particular advantage to the duodecimal scale and, for the operation of weighing and measuring capacities, the continual division by two rendered it practicable to make up any given quantity with the smallest number of standard weights and measures.

After completing their arguments on these points, the committee discussed the relative merits of the existing linear standards and recommended for adoption the standard yard employed by General Roy in the measurement of a base on Hounslow Heath. In the event that this standard was lost or damaged, it would be identified or recovered by gauging it according to the length of the mean solar seconds pendulum in London, at sea level, in a vacuum, and at 62° F. which was calculated as being 39.1372 inches on Roy's scale. Also, the French meter, representing the ten millionth part of the quadrantal arc of the meridian, was considered equal to 39.3694 of these inches.

The committee turned its attention next to measures of capacity and remarked that they were deduced most frequently from their relationship to measures of length. But since the quickest method of ascertaining the magnitude of any capacity measure was to weigh the quantity of water that it was capable of containing, it was thought advisable to adopt the latter approach. Using calculations advanced by Shuckburgh, the committee declared that 19 cubic inches of distilled water, at a temperature of 50° F., weighed exactly 10 troy ounces or 4800 grains, and that 7000 of these grains made an avoirdupois pound. These definitions were not advanced to introduce any departure from the contemporary standards of length and weight. But since the existing capacity measures did not agree in size or application, the committee proposed that they be equalized at one common capacity. After considering several alternatives, it was decided that the standard gallon should contain exactly 10 avoirdupois pounds of distilled water at 62° F., the entire vessel being nearly equal to 277.2 cubic inches.

In the second report the commissioners announced that they had discovered an error in the construction of some of the instruments employed in measuring the base on Hounslow Heath. Consequently, they rescinded their earlier recommendation and opted for the yard standard executed by Bird in 1760. They proposed that the Bird yard-bar be considered the foundation of all legal weights and measures, and they revised their calculations for the lengths of the London pendulum and the French meter to 39.13929 inches and 39.37079 inches respectively to conform to the new scale.

The third report presented the results of recent experimentation aimed at determining precisely the weight of a specific quantity of water. A cubic inch of distilled water weighed in a vacuum at 62° F. was now defined as 252.724 grains of the standard pound of 1758.[15]

Another committee was appointed by the House of Commons in 1821 to connsider the recommendations outlined in these reports. Agreeing with the general conclusions of their predecessors, and making only minor alterations, the new commissioners introduced a bill before Commons in 1823. A petition from the Chamber of Commerce at Glasgow to the House of Lords occasioned a further investigation and delayed the bill´s passage. Patrick Kelly (1756-1842), one of the witnesses before the committee and one of the most brilliant metrologists of the nineteenth century, called attention to the known effects of gravitational attraction on the pendulum, as shown by Kater´s own observations, and to the insufficient manner in which the level of the sea was understood. The Lords adjourned the entire proceedings temporarily, but in 1824 these matters were considered resolved and an act instituting the imperial system of weights and measures was passed.

Before examining this new system of weights and measures, however, it is necessary to discuss contemporary scientific and metrological matters in France. That activity led eventually to the creation of the metric system that would, in time, prove vastly superior to the imperial, and by the twentieth century would be adopted world-wide by virtually all industrialized nations. Most of the significant events in the metrological history of the British Isles during the later nineteenth and twentieth centuries would consist of a furious struggle between these two vastly different metrologies. In the end, the French scheme would win out.

4

BEGINNING OF REVOLUTION IN FRANCE: THE FORMATIVE YEARS OF RADICAL REFORM

During the Age of Science and the Enlightenment, France was plagued with the same metrological chaos as witnessed in the British Isles. There was a multitude of regional and local units owing to the fierce provincialism of duchies, counties, royal and aristocratic estates, cities, manors, and the like.[1] On the eve of the Revolution in the last quarter of the eighteenth century, France had more than 1000 units of measurement accepted as standards in Paris and the provinces, with approximately 250,000 local variations. Unit names were superfluous and confusing, and state standards bore little relationship to one another. Hundreds of ambiguous and misleading weights and measures decrees and laws had been issued from the initial reforms of Charlemagne (r. 771-814) in the late eighth and early ninth centuries.[2] A virtual army of inspectors, verifiers, and enforcers existed to administer the sundry laws and standards, but they were no more efficient, qualified, and free from graft and corruption than their British counterparts.[3] Control over weights and measures belonged to feudal lords as part of their fiefdom rights and, since taxes were based on units of measure, it was to the best interests of each lord to establish his own system and standards. During this era, however, various ministries labored conscientiously to reform the most serious problems of the metrology of the ancient regime by trying to force city and provincial governing bodies to adopt the system and standards of Paris. Embarking on a revisionist policy of its own, Paris envisioned a much smaller, more efficient, and more manageable system of weights and measures. Even though the government enjoyed some success, the eventual failure of revisionism, here as in England, was the spark that lit the fires of radical metric reform during the late eighteenth and early nineteenth centuries.[4]

Pre-Metric Standards

France had many Parisian standards for length, capacity, and weight that were considered by the government to be national prototypes, and it was upon these that the policy of revisionism was framed. First, the principal unit of length was the pied de roi (royal foot of 0.325 m) whose standard was represented by the legal toise (fathom) attached to a wall at the Grand Châtelet; hence its popular name of "Toise du Châtelet."[5] Also called the "ancient fathom of the masons," it was divided into 6 feet (1.949 m), each foot of 12 pouces (inches), each inch of 12 lignes (lines). Its length was marked by a rectangular step near each end of the bar, leaving the remainder at the ends half the thickness of the measuring portion.[6] This crudely constructed standard was altered several times after 1688, even by hammering to straighten it, the last such change occuring in 1766 when it was made equal to another line bar called the "Toise du Pérou" or Peruvian fathom. The latter standard, sometimes labeled the "Toise de l'Académie" or academician fathom because of its depository site, was constructed in 1735 by Charles Langlois, the royal engineer of Louis XV (r. 1715-74), the entire operation being supervised by Louis Godin (1704-60), a renowned astronomer. In May of 1766, Louis XV accepted this standard as the national prototype of linear measure replacing the Toise du Châtelet, and instructed the academician Tillet to produce 80 copies of it and to distribute them among the most important urban locations both in France and abroad. This fathom, together with another called the "Toise du Nord" or northern fathom, was used from 1735 to 1737 by astronomers Pierre Bouguer (1698-1758), Charles Marie de la Condamine (1701-74), and Godin in measuring an arc of the earth's meridian at Quito on the equator, and by Pierre Louis de Maupertuis (1698-1759) and Alexis Claude Clairaut (1713-65) near the North Pole in Lapland. The results of these measurements helped to prove Isaac Newton's (1642-1727) theory that the earth was an oblate sphere. Eventually serving as the model used to establish the value of the provisional meter of 1793 and the definitive meter of 1799, the Peruvian fathom was a flat bar of forged polished iron, about 1.5 inches in breadth and 0.33 inch in thickness; it was divided into feet, inches, and lines just as the Toise du Chatelet.

The last significant linear standard was the aune (ell) for cloth measurement maintained and supervised by the Marchands Merciers or Master Haberdashers in their guildhall. Fashioned after most toise standards, it was an iron bar with hooks or square projections at each end and engraved on the back "Aune des Marchands Merciers et Grossiers 1554." Measuring by statutory definition 3 feet, 7 inches, 8 lines (118.84 cm), its irregular subdivisions consisted mainly of halves, quarters, thirds, and sixths.[7]

Weight standards were preserved at the Hôtel des Monnaies or Royal Mint. From approximately the fifteenth century until the adoption of the metric system in August 1793, the official system of weights employed in France was called the "poids de marc" or mark weights, having for its primary standards the "pile de Charlemagne." Constructed sometime between the middle of the fourteenth and the end of the fifteenth centuries, the principal weight—the "livre poids de marc" or pound by mark weight of 0.4895 kilograms—was one of a series of 13 copper cup weights fitted one into another that totaled 50 marks or 25 pounds (12.2375 kg).[8] The mark served as the weight standard for French currency and for all transactions in precious metals. The pile de Charlemagne was later used by metric committees to determine the value of the provisional and definitive kilograms of 1795 and 1799 respectively. Even though there were discrepancies above and below the statutory grain allotments for several of the weights (especially those of 20, 14, 8, 4, and 1 marks), in determining the relation of the mark weight to metric weights, the committees regarded the entire pile de Charlemagne as a standard 50 statutory marks.

These linear and weight standards were generally constructed in adequate fashion and were reliable for the most part, despite the flaws and questionable practices mentioned. The same cannot be said, however, for French capacity standards because, just as in the British Isles, there were far too many of them, their accuracy was always suspect, and they were deposited at various scattered locations under the custody of competing individuals and groups. To cite only two examples, liquid standards were retained at the Hôtel de Ville under the care of the Provost of Merchants and Magistrates, while grain standards were conserved by the Master Salt

Measurers in their rooms at the same location. Originally these numerous prototypes differed according to whether they were employed for wheat, oats, salt, coal, and other products. Gradually reduced somewhat in number during the seventeenth and eighteenth centuries, the master measurers used them to verify local merchants´ measures.

Science, Technology, and Decimals

As mentioned in earlier chapters, metrological standardization was one of the major concerns of seventeenth century scientists. Their interest was spawned by the work around 1581 of Galileo on the pendulum, and his principle that the time of oscillation or vibration, the alternate ascents and descents, of a suspended weight is virtually independent of the amplitude or arc of its swing from any point on one side to the other. This discovery inspired them with the idea of basing a natural, universal standard on a length of a pendulum beating seconds at some predetermined location. A number of experiments began to select an ideal length, and to determine how that length varied according to the latitude of the place of observation. Among the most noteworthy advocates of a pendulum standard (despite their inability to agree on a common location based largely on national prejudices) were the Flemish engineer Isaac Beeckman (1588-1637), the Italian scientist Tito Livio Burattini (fl. 1600-40), the Dutch genius Huygens, and the French savants Picard, Gabriel Mouton (fl. 1640-80), and Marin Mersenne (1588-1648). They criticized severely the measuring units of Europe that made international and intranational comparisons very difficult to achieve and, out of desperation, they even attempted to solve this impass by reproducing the lengths of various European linear standards on paper by a common line graph. Although these reproductions were imprecise, notably due to the contraction of the paper over time, given the lack of availability of standards from one country to another, the line graphs served, at least temporarily, a useful purpose.

Perhaps based more on myth than historical fact, it has long been believed that Galileo discovered the principle of the pendulum by observing the swings of a large candelabra suspended from the ceiling in a Pisan cathedral. This swaying lamp was obviously only a crude approximation of a pendulum, but it is a well-established fact that Galileo later made more accurate experiments with a simple pendulum that consisted of a heavy ball or bob suspended by a cord. In an experiment designed to confirm his ideas of falling bodies, he lifted the ball to the level of a given altitude and released it. The ball ascended to the same height on the opposite side and confirmed his initial predictions. Later he postulated that the time of vibration or swing of this pendulum was directly proportional to the square root of its length, or inversely proportional to its acceleration due to gravity. He eventually used the pendulum to measure lapses of time, and designed a primitive model for a pendulum clock.

Slightly more than fifty years later, Mersenne made the first determination of the length of a seconds pendulum, and he published his results in Paris in 1644 under the title of <u>Cogitata physica: mathematica</u>. Following this scientific breakthrough, he proposed the problem of establishing the length of a simple pendulum equal in period of oscillation to a compound pendulum. The dilemma was not resolved by Mersenne but by Huygens, who, in his famous monograph of 1673, Horologium oscillatorum, sive de motu pendulorum ad horologia adaptato demonstrationes geometricae, established the theory of a compound pendulum. Huygens derived a theorem that a compound pendulum possesses conjugate points on opposite sides of the center of gravity; about these points, the time intervals of oscillations are the same. For each of these points as center of suspension the other point is the center of oscillation, and the distance between them is the length of the equivalent simple pendulum. Huygen´s work here was of monumental importance for the eventual perfection of the pendulum clock by Ahasuerus Fromanteel (fl. 1650-85), William Clement (fl. 1677-99), Joseph Knibb (ca. 1650-ca. 1711), John Harrison (1693-1776), Thomas Tompion (1639-1713), Graham, and other celebrated horologists. Harrison won fame for his gridiron pendulum and Graham for his mercurial pendulum.

These seventeenth-and early eighteenth-century scientists saw the pendulum as a possible standard for establishing an interrelationship between linear measurement (the primary standard), volume or capacity measurement, and weight or mass, by reference to a single standard for linear measures, that could be converted to standard volume, and then for weight could be converted to water, the principles upon which the later metric system would be based initially. At first they thought that they could introduce time into this sequence, but this was eschewed at this early date since a free-swinging pendulum of constant length would not keep the same time in different parts of the world, owing to the flattening of the poles originally discovered by Newton. (Later on the revolutionary committees would have further problems with time, but not for these scientific reasons.)

Other scientists, however, doubted the ultimate feasibility of a pendulum standard for weights and measures, and continued to espose one based upon a fraction of a great circle of the earth. These scientists were interested in the shape and, consequently, the measurement of the earth in topographical terms, and in the creation of geographical charts indispensable to navigation. One of the first scientists to attempt even a primitive geodetic operation was Jean Fernel (1497-1558), who in 1525 measured the meridian arc linking Paris and Amiens. To achieve this goal, Fernel counted the number of revolutions made by one of the wheels of a carriage in which he traveled over this route. At each revolution, a small bell rang inside the carriage. Taking into account the distance covered with each revolution, he concluded that the arc degree in question was equal to 56,746 fathoms and 4 feet. From such a crude beginning, great strides would be made during the next century.

Proceeding on the assumption that the earth was virtually a spheroid, seventeenth-century scientists maintained that it was possible to measure the arc of a great circle without much difficulty and without the awkward Fernel procedure. Such a measurement involved determining the latitude and longitude of two points, and then conducting a geodetic or trigonometrical survey that took into consideration the curvature of the earth´s surface, measuring the actual distance between them in terms of a unit of length selected for that purpose, and represented by a standard employed in the measurement of the baseline. The distance, as found by the triangulation, could then be compared with the difference in latitude between the two points, and the actual distance could be obtained in terms of the selected linear standard.[9]

But even before most of these pendulum and geodetic experiments had produced tangible results for metrological application, some scientists and technologists were creating ingenious methods for utilizing decimal notations for weights, measures, and money. Perhaps the earliest such scheme was one produced by a German assayer named Schreittmann (fl. 1560-90). In 1578 he devised a decimal system for the fractional weights—sixteenths, twelfths, quarters, and thirds—involved in various assay systems. He started with weights smaller than those used on assay-balances, and his primary unit, considered equal to 10 of these smaller weights, he labeled "elementlin oder atomi, stüplin oder minutslin." From this basic unit he built a scale consisting of 20, 30, 40, 100, 200, 300, 400, 1000, 2000, 3000, 4000, 10,000, 20,000, 30,000, 40,000, 100,000, 200,000, 300,000, and 400,000. He gave examples of computing assays in various systems by his decimal plan.[10] How great an impact this new method had in his own day is not known, but it is important as a stepping-stone to future endeavors.

Almost a decade later a far more important and comprehensive decimal scheme was devised by Simon Stevin (or Stevinus) (1548-1620), a Dutch mathematician, inspector of dikes, metrologist, and hydraulic engineer. In a monograph of 1585 entitled L'Arithmétique de Simon Stevin de Bruges, he established a scale to convert all fractions to decimals.[11] This method, that he called "disme," had a primary unit or whole number named "unité" (unity) that made 10 primes, the prime, 10 seconds, the second, 10 thirds, the third, 10 fourths, and so on. The signs designating the decimals were (0) = unity, (1) = prime, (2) = second, (3) = third, (4) = fourth, and the like, and they were placed over the appropriate intergers (herein they are recorded elevated to the right of the intergers). To transcribe a fraction such as 77 23/100, for example, he wrote $7\ 7^{(0)}\ 2^{(2)}\ 3^{(3)}$, or a fraction such as 4/100 became $4^{(3)}$.

This awkward system that employed superscripted decimal figures was modified substantially during the seventeenth century by mathematicians Bartholomew Pitiscus of Grunberg (1561-1613), John Napier (1550-1617), Jacques Hume (fl. 1610-40), Albert Girard (1595-1632), and André Tacquet (1612-60). The first to simplify Stevin's decimal notations was the Dutch scholar Pitiscus. In the third edition of his trigonometric work—Silesii Trigonometriae, sive de dimensione triangulorum, libre quinque—published in Frankfort in 1612, he advocated eliminating Stevin's cumbersome superscripted decimals in favor of a shorter, more convenient method. Now, instead of writing 4/100 as $4^{(3)}$, it was easier to transcribe it as 004; or 9/10, in the Stevin system = $9^{(2)}$, was simply 09.

Within this same decade, the Frenchman Napier, the inventor of logarithms, proposed in his famous work—Mirifici logarithmorum canonis constructio et eorum ad naturales ipsorum numeros habitudines—that fractions such as 10000000.04 and 10000000 4/100 were one and the same number expressed in different fashions. Thus, he became the first scientist to incorporate the point or period to designate decimals that is still the most common method employed today.

In 1636 the Swiss Hume, in his Traité de la Trigonométrie, employed a zero to replace the primes, seconds, thirds, etc. of Stevin's system (zero elevated to the right); hence, 2 81/100 became $2^{(0)}8\ 1(3)$.

Girard, a contemporary of Hume, represented a slight step backwards, since he advocated eliminating the points of Napier and the zeroes of Hume and opted for Roman capitals instead. In Girard's system the figure 7.653 was 7653 III, while 0.000001818 became 1818 IX.

Finally, the French Jesuit Tacquet, in an arithmetical work published at Anvers in 1665, incorporated both the Stevin and Girard systems; therefore, the fraction 2 81/100 was transcribed as (once again the appropriate signs are elevated to the right rather than above the intergers) $2\ 8^{II}\ 1^{III}$.

Over the course of the eighteenth century, European scientists eliminated all of these proposals except one. It was Napier's system that was adopted universally by the time that the metric system and its decimalized metrology became a reality during the 1790s. The conjunction was fortuitous indeed.

Pre-Metric Reform Proposals

The earliest complete plan proposed to alter significantly the metrology of France, and to provide more reliable standards based on some of the scientific findings and decimal schemes discussed above, was that of Gabriel Mouton, mathematician, astronomer, and vicar of the collegiate church of St. Paul in Lyons. Prior to his published monograph of 1670, and motivated by his desire to prevent the loss or alteration of the standard of the Toise du Châtelet that recently had been issued, he advocated fixing its true length either on a seconds pendulum or on a minute of a terrestrial degree.[12] Eventually selecting both of these scientific methods in various ways for establishing an immutable, universal standard, and becoming the first distinguished scientist to espouse a system of weights and measures based upon decimal numeration, his novel plan initially derived its primary unit (milliare) from a minute of the arc of the largest circle that could be drawn around the world, either a full line of longitude or latitude.[13] Taking a minute, or 1/60 of a terrestrial degree, the largest unit, he divided this as yet undetermined magnitude into 10 million parts or points, and developed the following scale. A minute of arc was called a milliare or mille; one-tenth of a milliare, a stadium or stade; one-tenth stadium or 1/100 milliare, a funiculus, chaine, or corde; one-tenth funiculus or 1/1000 milliare, a virga or verge; one-tenth virga or 1/10,000 milliare, a virgula or small verge; one-tenth virgula or 1/100,000 milliare, a digitus or doigt; one-tenth digitus or 1/1,000,000 milliare, a granum or grain; and one-tenth granum or 1/10,000,000 milliare, a punctum or point. These terms could be used interchangeably with a second set consisting of milliare, centuria, decuria, virga, virgula, decima, centesima, and millesima. In tabular form, Mouton´s system before 1665 was as follows.

Table 4.1

Relationship of Units in Mouton´s Plan

millesima	centesima	decima	virgula	virga	decuria	centuria	milliare
10							
100	10						
1000	100	10					
10,000	1000	100	10				
100,000	10,000	1000	100	10			
1,000,000	100,000	10,000	1000	100	10		
10,000,000	1,000,000	100,000	10,000	1000	100	10	1

In this proposal there were two major measuring units, the virga for large measures and the virgula for shorter ones. They corresponded roughly to the toise and pied of the old system.

Just when this plan was proposed is not known precisely, but it could not have taken place before 1665 since some observations in connection with the fixing of his standards were not made until March of that year. Unfortunately, geodesy had not yet reached a stage when its results were immediately applicable to metrological reform. In fact, only a few determinations of the length of terrestrial degrees had been made by this time, those of Snell, Richard Norwood (ca. 1590-1675), and Giovanni Battista Riccioli (1598-1671) being the most important. Snell, a Dutch mathematician, cartographer, surveyor, and professor at Leiden, introduced trigonometric methods in his operations, so that instead of measuring the entire distance of an arc, he only measured a short base, and by means of a system of triangles, he computed the distance between the two extreme places. Near Leiden, for example, he measured a base of 326.43 Rhineland perches, this perch equal to 12 Rhineland feet. Working from this, he established geometrically the relation between the principal cities of Holland and several in Flanders, extending his work as far north as Alkmaar, near Leiden, and as far south as Bergen-op-Zoom. The latitude of these three places was observed from stars. The differences in latitude between Alkmaar and Bergen-op-Zoom was $1^{\circ}11'30''$, and the meridian distance corresponding to the latitude stations was 33,930 perches, giving as a value of 1 degree, 28,473 perches. By using Alkmaar and Leiden, the length of one degree was found to be 28,510 perches; the round number 28,500 perches being selected as the probable value. Converted by line graphs to the French system, Snell computed the length of one degree to be the equivalent of 55,100 French fathoms.

During the years 1633-1635, the English scientist Norwood made an arc measurement between London and York. The results published in 1636 found a difference of latitude between these two cities of $2^{\circ}28'$, and he found the length of one degree therein to be 57,442 fathoms.

Finally, Riccioli conducted a trigonometric survey in Lombardy around

1650. He measured a base near Bologna of 1094 Bolognese paces plus 2 1/4 feet, and from a chain of triangles, he obtained the distance between Bologna and Modena. Instead of reducing to meridian distances as earlier surveyors had done, he tried to find the arc of the great circle passing through the two places, from the zenith observation of stars. The results of the value of one degree varied from 56,130 to 61,797 fathoms. Later he measured vertical angles to the different stations and obtained values of approximately 62,650 fathoms for one degree. The least acceptable of all methods of triangulation, the idea here was to measure upon any ground surface the largest linear distance possible, and then to calculate the angles of the two respective verticals with the common line of vision comprised between the extremes of the base.

It is regrettable that Mouton appears to have been unaware of the results of Snell and Norwood for he relied completely on the findings of Riccioli, the least accurate of the various surveying attempts. In time several scientists showed that Riccioli´s base was too short, that only two angles of each triangle were observed, that many of the angles were too small, that some were determined indirectly as sums or differences of other angles, that no correlations were made for refraction, and that some distances were estimated from meandered lines. Snell´s earlier triangulation results were far superior; Mouton´s metrological system would have benefitted by their incorporation.

Riccioli gave his degree length in terms of the Bolognese foot and the old Roman foot. Since he found a degree to be 321,815 Bolognese feet,

Mouton calculated his virgula at 6.44 of these feet. Even though the pendulum principle was discovered only a few years before this, and knowing that accurate measurements of one minute of a meridian arc was a terribly difficult task, he chose, as a backup, the seconds pendulum as the best means of preserving and transmitting the virgula standard, making it equal to a pendulum that beat 3,959.2 vibrations in a half hour at Lyons. He performed experiments on a pendulum 1.852 millimeters long (the virga) in order to determine the number of oscillations in a half hour, and the average of his observations varied between 1,251.8 and 1,252.1 oscillations, a remarkable result since theoretically this number should be around 1,252.26. For the virgula, the tenth part of the virga, the theoretical value was 3,960. Mouton found an average of 3,959.2. Once this linear standard was defined, he maintained that the units of capacity, volume, and weight could be easily derived from it although he never stated how or provided their names and relationships. (As seen earlier, this failure hampered some of the English reform proposals as well; new plans needed completion to have a noticeable impact.) In later metric terms, the values of his linear units can be expressed as follows.

Table 4.2

Metric Values of Mouton's Linear Units

Unit	Designation	Metric Value
milliare	1 minute of arc	1,855.3 m
centuria	1/10 milliare	185.53 m
decuria	1/10 centuria	18.553 m
virga[a]	1/10 decuria	1.8553 m
virgula	1/10 virga	18.55 cm
decima	1/10 virgula	1.855 cm
centesima	1/10 decima	1.855 mm
millesima	1/10 centesima	0.1855 mm

[a]The virga, conveniently in the middle, was closely approximate to the Parisian fathom of 1.949 meters.

Mouton published these results in Lyons in 1670 under the title—*Novae mensurarum geometricarum idea et novus methodus eas et quascumque alias mensuras communicandi*. The work established Mouton as a metrological pioneer, the first significant pre-metric French reformer, for three reasons: he devised a system based upon a decimal scale; he derived his primary unit from the length of a minute of the terrestrial arc; and he showed how this unit could be expressed in terms of the seconds pendulum. To be sure, his system was incomplete, but he was responsible for taking the first full step on the road to the metric solution.

From the publication of Mouton's pioneering plan in 1670, through the period encompassed by Picard's and Cassini's far less significant reform attempts of 1671 and 1720, the pace of pendulum and geodetical experimentation quickened. Motivated to a great degree by Mouton's researches, a number of European scientists added substantially to the operation and effectiveness of these two methods of establishing a universal standard. The earliest of these was Picard, an eminent astronomer and geodesist. Having worked on a restoration of the Toise du Châtelet in 1666, he proposed in his famous book—La Mesure de la terre—published in 1671, that it was imperative that this newly restored fathom be based on an invariable standard taken from nature. To furnish this he undertook a measurement of the arc between Malvoisine, just south of Paris, and Sourdon, near Amiens. Using a new reticular eyepiece or telescope with spider lines that improved considerably the determination of the true length of angles, he measured two bases. His first, lying between Villejuif and Juvisi, consisted of 5,663 fathoms; the second, near Sourdon, of 3,902 fathoms. A chain of 13 triangles was formed. The distance included between the ends of the meridian arc came to 78,850 fathoms, and the difference between the latitudes being $1^{\circ}22'5''$, the arc degree came to 57,060 fathoms, considered the mean between his final measurements of 57,064 and 57,057 fathoms. Despite some errors that were discovered later on, Newton employed this distance to check the theoretical investigation of his law of attraction, and eighteenth-century scientists considered

Picard´s work to be the real starting point for accurate geodetic determinations, replacing those of his predecessors.

Picard also made a contribution to pendulum research when he discovered in 1669 that the rates of clock pendulums changed whenever their lengths were altered by expansion and contraction due to increases or decreases in temperature. (This would lead later to a revolution in the construction of pendulum rods that utilized bi-metals.) Unfortunately, he used these pendulum experiments two years later to propose an extremely underdeveloped reform plan that had little chance of success and, unlike Mouton´s proposal, was quickly ignored. Stating that the length of some permanently established pendulum could constitute a universal standard that he labeled "the astronomical radius," he believed that a third of its length could be used to establish a "universal foot," its double, a "universal fathom," its quadruple, a "universal rod," and 1000 of these rods, a "universal mile." The major problem here, of course, was that the standard could only apply to the immediate site where it was determined and housed. It could never be duplicated anywhere else in the world. This, coupled with the obvious lack of completeness, doomed his metrological efforts.

Regardless, Picard´s important geodetic work made a distinct and permanent impression. Upon the insistance of the Academy, Louis XV appointed a committee chaired by Cassini and de la Hire to verify Picard´s geodetic findings and to extend its range. Interrupted in 1683 by the death of the Academy´s protector, it did not recommence until 1700. During the next eighteen years the work was completed by Giacomo Cassini (1677-1756) who had succeeded to the position of royal astronomer upon the death of his father.[14] The triangulation was extended south to Spain, and near Perpignan, along the shore of the gulf of Lyons, a base line of verification was measured. Encountering many problems such as fogs near shore lines, a frequently broken base line, water and sand dune interference, instrument errors, and occasional local resistance (a problem that would plague French scientists later engaged in determining the geodetic measurements upon which the metric system would depend), the observations at the southern extremity were made at Collimere in March, 1701, while the northern observations took place at Paris in March, 1702, the same star being used at both locations. The meridian amplitude, allowing for refraction and change in declination, was found to be $6^{\circ}18'57''$, and the completed distance between the astronomical stations, when reduced to sea level, gave a value of 57,097 fathoms for one degree. This was greater than Picard's value by 37 fathoms, and produced further evidence that the earth was elongated at the poles.

The triangulation was then extended northward from Paris. Using nine of the triangles that Picard had established, and adding twenty more, the scientific team extended the line to Dunkirk. At the extremity near Dunkirk, a northern verification base was measured along the seashore. Some 4000 fathoms of the base were remeasured, giving a difference of three feet from the first measurement. The final value of the base line was 5564 fathoms. The computed value from the triangulation agreed within one fathom. The star observations were made at Dunkirk in 1718, using the same type of sector that Picard had used. The amplitude of the meridian arc was determined to be $2^{\circ}12'9.3''$, and the corresponding distance was 125,454 fathoms, making one degree equal to 56,960 fathoms, 100 fathoms smaller than Picard's value. This reinforced further the substance of Newton's theory.[15]

On the basis of these experiments, and similar at least in design to the plan of Mouton, Giacomo Cassini in 1720, under the title of De la grandeur et de la figure de la Terre, recommended the adoption of a unit called the "geometrical foot" (pied geométrique). This was equal to 1/6000 part of a minute of arc of the great circle that he had measured, and six of these feet made his new standard for the fathom. Never expanding this primary unit to include other linear units, or their proportional relationships, or even applying it to other measurement divisions such as capacity measures or weights, spelled, like other incomplete European plans before his, its almost immediate demise. Like Picard, he had failed to do more than suggest a universal standard. Admittedly his choice was better, but like his illustrious predecessor, he could only suggest an ideal system of weights and measures, he could not create one.

No more successful than the Picard and Cassini metrological reform plans was one offered by Condamine in 1747 in a work entitled Nouveau Projet d'une mesure invariable propre à servir de mesure commune à toutes les nations. Rejecting the reliability of the academician fathom as a universal standard, he urged that one be adopted based on a pendulum beating seconds at the equator. What is unusual about this selection is that it followed his participation in one of the most successful geodetic operations conducted up to that time. Starting in 1735, he, together with Godin and Bouguer, were commissioned to measure an arc near the equator, while others, such as Maupertuis, Réginald Outhier (1694-1774), Charles Etienne Louis Camus (1699-1768), Clairaut, Pierre LeMonnier (1675-1757), and the Swedish astronomer Anders Celsius (1701-44), were sent to Lapland. After years of weather-related ordeals both north and south, they proved that the length of the meridian arc increased with increasing latitude. In Bouguer's published result—La Figure de la terre—the value of the ellipticity of the earth was derived from a comparison of 3 arcs, using duplicates of the Peruvian and northern fathoms. The first was in Peru at 2° south latitude; the second in France at approximately 40° north latitude; the last in Lapland at 60° north latitude. In the northern triangulations the committee members derived a value of one degree of 57,437.9 fathoms, or 378 fathoms greater than Picard's value, a figure later shown to be too large. The southern triangulations were divided among three separate groups, and they produced calculations of 56,746, 56,749, and 56,768 fathoms, the figure agreed upon being 56,753. The accuracy of Newton's theoretical calculations of the polar compression of the terrestrial globe became fact. (As of this date, this was perhaps the most important contribution of metrological research to scientific evolution.)

Just when Condamine found time to conduct pendulum experiments is not known, but he did establish a length of a seconds pendulum at the equator of 36 inches, 7.15 lines based on the Peruvian fathom. He also praised the advantages of a decimal division of weights and measures, one begun with linear measures, with the others brought into harmony in a stable and invariable fashion. But, aside from advocating another pendulum possibility for a universal standard, he offered little else, and his reform proposal died as rapidly as those of Picard and Cassini. Like Picard, he would be remembered for his geodetic contributions, not for his metrological refinements.

There were other reformers in France besides these. In addition to adding substantially to pendulum and meridian arc refinements, they proposed such possible adoptions as platinum rods to replace earlier ineffective prototype standards, and they even proposed a reformation of the currency along decimal lines that one scientist labeled decimes (tenths) and centimes (hundredths). Both of these suggestions would eventually become important segments of the metric system, the latter still part of the French monetary operation. Their overall plans failed for many of the same reasons analyzed above, but metrication was approaching. Over the next decade, the foundation for the best metrological system ever conceived and brought to fruition would become a reality.

Dawn of Metrics

The impact of this monumental work during the seventeenth and eighteenth centuries by scientists from many European nations stirred the French government into action. The only remedy that appeared feasible to

French scientists was a complete renunciation of the weights and measures of the ancient regime and the creation of a totally new system that would be precise, simple, and coherent. A political revolution was imminent; France's new metrology would reflect this revolutionary experience.[16]

Before the first revolutionary metrological proposal in 1790, however, and beginning early in the reign of Louis XVI (1774-92), Turgot, Comptroller General of Finances, offered one last pre-metric reform plan. Opting once again for a seconds pendulum standardized system at 45° latitude, because Charles Messier (1730-1817) had established a value for it in 1775, and because Jean Antoine Condorcet (1743-94) had drafted an instruction to this effect, the project failed, perhaps luckily for the future of metrication, when the political situation deteriorated badly and when Turgot was replaced. The metric system was in the wings as assemblies of baillages reunited in 1789 for the election of deputies to the Estates General. They demanded uniformity of weights and measures (motivated, perhaps, more by political and social reasons than by scientific concerns, since anything associated with past repressions was considered ripe for change), and an end to a thousand years of metrological confusion.[17] Eleven years earlier Jacques Necker (1732-1804), in a report made to Louis XVI, had pointed out the difficulties arising from a radical restructuring of weights and measures, among which the most damaging was the mass confusion that would be caused in financial contracts, yearly rent payments, feudal rights, legislation, and various types of printed materials.[18] The activities of the 1790s in France constituted an outright rejection of Necker´s fears. (Given the turbulent political situation, Necker was worried more, perhaps, by the political ramifications than by the metrological obstacles.) A similar situation in Great Britain would not occur until the second half of the twentieth century.

The first major stride was initiated by the Bishop of Autun, Charles Maurice de Talleyrand-Périgord (1754-1838), who submitted a proposal in April 1790 entitled Proposition faite à l´Assemblée Nationale, sur les poids et mesures. Talleyrand, to say the least, had a checkered political career. Swaying with the political winds from a representative of the clergy in the Estates General to opposing eventually their inclusion in the National Assembly, he became an advocate of the radical reforms sweeping France after 1790. Renouncing his bishopric, excommunicated by the Pope, speaking in favor of free education all the way to the university level, this "radical" Talleyrand, after having recognized the value of the major reform projects since Mouton, dismissed Necker´s partial reform scheme of 1788 since the old problems were not resolved. With the far-flung changes being wrought by the Revolution in the political structure, he believed that it was possible to gain a receptive audience for massive metrological change. He referred specifically to a situation in which the revolutionaries produced a new flag, abolished the old royal titles along with the absolutist monarchy, promulgated a Declaration of the Rights of Man, did away with the old provinces and governments, established a "New France" of 83 departments or electoral precincts, and confiscated royal and ecclesiastical estates. It is within this turbulent climate that

Talleyrand argued vociferously for a new, uniform system of weights and measures with no ties to the ancient regime. He wanted a system that was international in scope and application; one based upon the latest scientific discoveries to facilitate its adoption among all nations.

Talleyrand did not provide the names or unit proportions for any weights and measures; rather he pleaded in his report for some invariable standard drawn from nature such as a seconds pendulum at 45° latitude; a length to be determined with exact precision by new observations. For weight he advocated a cube of water whose height was one-twelfth the length of the pendulum.

An integral part of this report was Talleyrand's request that England collaborate in the venture, and he proposed the formation of a joint commission to consist of members of the Academy and the Royal Society. To expedite its resolution, he sent a private letter and a printed copy of his plan to Riggs Miller, the English metrologist, who had earlier proposed a system based upon the length of a seconds pendulum.[19] An added inducement was the fact that in the United States, Thomas Jefferson, the Secretary of State, was advocating a pendulum standard established at 38° latitude, the mean latitude of the United States.[20]

Talleyrand's plan for joint sponsorship of a new system based upon French, English, and American imput failed. For six months England did not respond. On December 3, 1790, shortly after the Miller project was placed before the House of Commons (April 13, 1790), the Duke of Leeds, Secretary for Foreign Affairs, replied to the Marquis de la Luzerne, the French

Ambassador, that the British government regretted that it could not accept Talleyrand´s proposal, the arrangement being considered "impracticable." The English feared the radical changes going on across the Channel, and probably reasoned that cooperation on this metrological venture would give tacit approval to republican political upheavals and bloodshed.

In the United States, President George Washington (1732-99) on January 8, 1790, sent a message to Congress inviting them to study America´s system of weights and measures. In this presidential address, the first ever to Congress, Washington informed the assembled representatives that, as in England, there was no metrological uniformity throughout the States of the Union, and that the continued proliferation of local and state units of measurement would have detrimental consequences in the future. America had inherited the weights and measures of Great Britain that, by the later stages of the Industrial Revolution, had grown to slightly more than 500 national standard units having more than 100,000 local variations. Washington knew that such proliferation, although not yet a reality in America, could become a definite possibility if immediate steps were not initiated. He also informed Congress that the Constitution mandated the representatives of both Houses to concern themselves with the task of metrological uniformity.

Several weeks following this address, a select committee of Congress was appointed to handle the problem. Especially important in the ensuing deliberations was the inspiration, work, and direction supplied by Jefferson who, amazingly, is little known for his considerable pioneering achievements on behalf of metrological decimalization. Being one of the dominant personalities of the early American intellectual scene, and a valuable friend, confidant, and ally of many Enlightenment savants of both England and France, Jefferson saw two ways of attacking the problem.

First, he believed that it would be possible to achieve a degree of metrological uniformity simply by reforming the existing system of measurement. This plan entailed the following specific recommendations: (1) the elimination of all superfluous and outdated national units and the total eradication of local units employed in contravention of government directives; (2) the manufacture and dissemination of a new set of national standards that would be used to size and authenticate weights and measures employed in all economic and social sectors; (3) the simplification and realignment of all multiples and submultiples of measurement to eliminate the present confusing and overly complex system that randomly, and somewhat irrationally, contained binary, sexagesimal, duodecimal, and other divisions; and (4) the complete restructuring of acceptable measurement units based upon some invariable natural standard, such as one predicated on earth measurements or on a pendulum beating seconds at an agreed-upon altitude. If all of these criteria could be researched properly, organized competently, and enacted nationally into law, the United States would achieve at least a partial solution to its metrological problems.[21]

Jefferson's second proposal, however, was far more revolutionary, for it advocated the elimination of the entire English pre-imperial system and the adoption of a decimal system analogous to the one that the revolutionary government of France began to work on in earnest following the storming of the Bastille in July of 1789. The new system would be based on the seconds pendulum, with decimal multiples and submultiples of the fundamental units, the primary unit equal to three-tenths of the pendulum's length.

Unfortunately for the future of both of these plans, and for the adoption of the French plan, Congress did not respond. It established a long-lasting trend for nothing was done on the weights and measures question during the administrations of the first five American presidents, even though Congress had been officially challenged to act on this problem on at least twelve occasions.[22]

Besides the obvious obstacles presented by more immediate and pressing matters that weighed heavily on the young government, American collaboration in foreign metrological reform was stymied for two specific reasons. Jefferson, though enthused with the possibility of an invariable, universal standard, could not accept a meridian-based French standard. He believed that the base of any new system should be one readily verified in all supporting nations. Since the proposed meridian measurement could not be made anywhere other than between Dunkirk and Barcelona, the United States would be at a decided disadvantage in any future verification corrections. (The same dilemma would be faced, of course, had France and other European states adopted Jefferson's recommendation for a United States location for a pendulum standard.) Also, when Jefferson heard of the English response, and since America had just recently won its independence from England, quite justifiably from a diplomatic perspective, he did not want to renew hostile relations with a recently vanquished foe. In the end France merely secured a statement of support from the Spanish government, due principally to the fact that Barcelona was one of the axis points along the selected meridian arc. Spain, however, did not participate actively in the actual operations leading to the establishment of the metric system. France had to embark on its metrological mission alone.[23] Joint collaboration by the American, British, and French governments at this juncture could have avoided in the long run the impass that plagued Great Britain until the 1960s and the United States to the present day. But it was not to be.[24]

While France awaited these responses, Talleyrand´s proposal was sent to the National Assembly´s Committee on Agriculture and Commerce that charged the president of the committee, the Marquis de Bonnay, with the task of compiling a report. The report was ready on May 6, 1790, and the discussions began on May 8. On the same day the Assembly issued a decree that conformed very closely to Talleyrand´s proposition, and stated that its primary object was to determine a natural standard upon which to base the new system. Surprizingly, it did not insist on the latitude of 45° while opting initially for the seconds pendulum as the standard of linear measures.[25] Since the English response was not yet in, the maneuver may have served as a diplomatic inducement for foreign cooperation. Perhaps for the same reason, the decree made no mention of a standard derived from a meridian arc or of a decimal scale of unit proportions. It did ask administrators of the newly formed departments to send prototypes of their standards to Paris where they would be retained under the supervision of the Academy. Certain academicians were ordered to compare these local prototypes with those of the proposed new system, issue books to all municipalities that compared their dimensions, distribute any new standards manufactured throughout the realm, and destroy the old standards within six months of the decree. The final instruction to the Academy was to indicate which scale of numeration—decimal, duodecimal, or any other—was the most convenient for weights, measures, and coins.

As a result of this decree, the Academy selected a committee consisting of Condorcet, Tillet, Jean Charles Borda (1733-99), a famous instrument designer and nautical astronomer who advanced the science of hydrodynamics and devised improvements in repeating and reflecting circles used in celestial and land surveying, Joseph Louis Lagrange (1736-1813), a celebrated mathematician and astronomer often compared in importance with Newton in the realm of analytic mechanics, and Pierre Simon, Marquis de Laplace (1749-1827), astronomer, geometer, government senator, and minister of the interior. On October 27, 1790, its report opted for a decimal system of weights, measures, and coins after some pronounced dissent. It considered a duodecimal system at some length before finally rejecting it, but one member expressed regret that the base of 10 could be divided evenly only by factors of 2 and 5, whereas the base of 12 could be divided by 2, 3, 4, and 6. Lagrange even advocated an 11-base system, not evenly divisible save by itself and 1. This was quickly dismissed by the other members as foolhardy.

Even though the report of this first committee answered many of the questions raised in its original directive, a second had to be appointed to settle the issue of the appropriate universal standard. Its members were Borda, Lagrange, Laplace, Condorcet, and Gaspard Monge (1746-1818), mathematician and descriptive geometer, who was one of the founders of the Ecole Polytechnique, professor of physics at Lyons at the age of 16, and president of the Egyptian Institute. Cooperating with this and other committees until he was guillotined during the Reign of Terror was Antoine Laurent Lavoisier (1743-94), the most famous chemist of the eighteenth century. In a report of March 19, 1791, entitled _Sur le choix d'une unité de mesures_, this new committee saw three possible choices for a standard. The first was the length of a seconds pendulum. This was considered easy to determine and to verify in case of accident. That of 45° was preferred because it was the mean length between the equator and the North Pole. Thus, it was the average length of pendulums beating seconds at diverse latitudes. However, as a unit of length it was not convenient since it was dependent on acceleration due to gravity, and it was necessary to have its position specified exactly. Even more damaging was the fact that the pendulum introduced a new element, the second of time, and this was considered a non-linear function and an arbitrary division of the day. For these reasons, the pendulum was eliminated as a possible model for an ideal standard.

The second choice—that of a quarter of a circle of the terrestrial equator—was also eschewed. Its regularity was no more assured than the regularity of the similitude of meridians. Also, equatorial regions would prove too difficult to gain access to, and there would be serious problems resulting from climatic and other conditions.

So the committee selected the quadrant of the terrestrial meridian. (An unusual choice since the committee members must have been aware of the enormous distance that would have to be measured and the inordinate amount of time that such an operation would consume.) After an arc had been measured, the length of a quadrant would then be computed, and one ten-millionth of its length would be taken as the base or primary unit of length, with all other units derived from it according to a decimal scale.[26] Also, the quadrant was to be measured in a single unit of length on a decimal basis instead of in the former degrees, minutes, and seconds of the old system. The committee decided further that the unit of length, once chosen, would be used to establish the units of area and volume, and the basic weight would be that of a designated volume of pure water. The linear unit—one ten-millionth of a quadrant of a meridian arc—would eventually be called the meter. The weight standard at this juncture would be the amount of water in a cube whose side was one-hundredth of the, as yet, undetermined meter.

Certain problems arose soon thereafter. The measure of a quadrant of the entire meridian was judged impractical. It would suffice to choose a sufficiently large arc whose extremities touch to the north and south of 45° parallel at distances that, if not exactly equal, were not too disproportionate. The arc chosen went from Dunkirk on the northern French coast to Barcelona on the Mediterranean Sea. Due to additional experimental evidence, these two aforementioned axis points were selected because they were situated at sea level in the same meridian, they afforded a suitable intervening distance of approximately $9^{\circ}30'$ (the largest in Europe available for a meridian measurement), the territory had previously been surveyed trigonometrically by Abbot Nicholas Louis de Lacaille (1713-62), a respected astronomer, and Giacomo Cassini in 1739-41, and such an arc extended on both sides of latitude 45°.

The report of March 19 set six goals: determine the difference of latitude between Dunkirk and Barcelona; measure the old bases; establish and measure the series of triangles used in a previous survey and extend the same from Dunkirk to Barcelona; observe pendulum differences at 45° latitude; verify the weight in vacuum of a known volume of distilled (replacing the term "pure") water at the temperature of melting ice; and compare the old measures with the new while constructing scales and tables of equalization values. This second report was adopted by the Academy and transmitted to the Assembly on March 26, 1791. It must be emphasized, however, that there was still a hard core of resistance from those scientists who preferred the pendulum standard, and from those who thought that the meridian measurement was already precise enough to establish a new system since it had been determined on two earlier occasions. (The opposition would later prove to be wrong.)

That same day, Talleyrand endorsed the project and sent copies of it to the Committee of Constitution, the Committee of Agriculture and Commerce, and the members of the Academy. The project was sanctioned four days later on March 30, 1791, the government ordering that France adopt the quarter of the meridian as the base of the new metrological system, and that the necessary operations, notably the measurement of the arc from Dunkirk to Barcelona, begin immediately.

The Academy divided the work among five commissions. The triangulation and determination of latitudes were given to Giacomo Cassini, Pierre François André Méchain (1744-1805), astronomer and discoverer of several comets whose orbits he calculated, and Adrian Marie Legendre (1752-1833); the measurement of the bases to Monge and General Jean Baptiste Marie Meusnier (1754-93); the length of the seconds pendulum to Borda and Charles Augustin Coulomb (1736-1806); the weight of a known volume of water to Lavoisier and René Juste Haüy (1743-1822); and the comparison of the provincial measures with those of Paris to Tillet, Mathurin Jacques Brisson (1723-1806), and M. Vandermonde (1735-96).[27] A central commission was also named to direct all of the operations, and it consisted of Borda, Condorcet, Lagrange, and Lavoisier.

On the first commission Legendre soon quit since he was not interested in measuring the triangles of the meridian; Cassini left soon thereafter to retire from the Observatory. These scientists were replaced by Jean Baptiste Joseph Delambre (1749-1822), who later held the chair of astronomy at the College of France. On the second commission, General Meusnier was called to serve on June 13, 1793 in the Army of the Rhine, leaving Monge to execute the measurement of the bases alone. But Monge appears to have done nothing, and the task of measuring the bases fell to Méchain and Delambre. Delambre alone made more than two-thirds of the angular determinations, the evaluation of the lengths of the two bases, and he assumed the task of calculating, of discussing, and of publishing the results. On the third commission Coulomb was replaced by Cassini, who claimed later that he made all of the observations, while Borda created the basic methodology to be employed. The most stable and efficient commission was the fourth, perhaps due to the fact that it entailed the least personal inconvenience resulting from traveling to distant parts. Lavoisier and Haüy, a mineralogist, had almost completed their work (save for determining the variations in volume and weight arising from different degrees of temperature) when Lavoisier died at the hands of the Revolution. Louis Lefèvre-Gineau (1751-1829) and Giovanni Fabbroni (1752-1822) eventually completed this project. On December 20, 1791, the fifth commission lost Tillet; Claude Louis Berthollet (1748-1822) was his replacement, but the basic project entrusted to them was never completed.

The First Metric System

During the spring of 1793 the Academy issued its recommendations to the government. First, it fixed the new linear standard, called the provisional meter, at 36 inches, 11.44 lines based upon the ten-millionth part of the quadrant of the meridian.[28] This standard was derived from a calculation of the observations made by Lacaille when measuring a meridian in France in 1740. This triangulation gave a value of one degree as 57,027 fathoms, that multiplied by 90 produced a length of 5,132,430 fathoms for the entire quadrant. Taking the ten-millionth part of this value, and reducing it into feet and lines, produced the value given for the meter of 3 feet, 11.44 lines. The Peruvian fathom was employed in the establishment of this value, and a standard of this provisional meter was constructed by Etienne Lenoir (1744-1832) in Paris.

The Academy further recommended that the new system of weights and measures be based on a complete decimal scale, and they even extended this to time and angle divisions. The latter moves, together with a later restructuring of the Gregorian calendar, constituted the only major significant failures during this dynamic decade of change. New decimalized clock units were introduced by the autumn of 1793, and they were met immediately with massive resistance. People rebelled against 10-hour days, 100-minute hours, and 100-second minutes, and found it extremely awkward to refer to a deciday (2.4 hours), milliday (86.4 seconds), and 10 microdays (0.864 second). The horologists of France shouted the loudest in opposition, since this ill-conceived plan rendered obsolete every watch and clock ever produced, either inside or outside of France. By 1795 this unfortunate "reform" was tabled and it was never reconsidered. The same fate awaited the new circle of 400 degrees, the quadrant of 100 degrees, the degree of 100 minutes, and the minute of space of 100 seconds. In these two aspects the Revolution had gone too far; radicalism had overstepped its boundaries. The calendar reform initiated later in 1793 suffered the same fate.

The final recommendations of the Academy established the nomenclature for the new system and the interrelationships that were to exist among linear and capacity measures, weights, and monies. The selection of appropriate unit names presented the greatest problem. Two systems emerged from the Academy´s deliberations: the "methodical," in which Latin prefixes were used in conjunction with the root-word meter, (later extended to include Greek prefixes as well), and the "simple," monosyllabic names that some believed would be more readily accepted by the general population.

The recommendations of the Academy were submitted to the National Convention that had replaced the Legislative Assembly in September, 1792, just as the latter had replaced the National Assembly in October, 1791. In the Convention, the Academy´s suggestions were discussed by the Committee of Public Instruction, and Abrogast, a radical leader, was instructed to bring the new system into operation. On the strength of his report, a decree of August 1, 1793, adopted the Academy´s general recommendations, ruled that the "methodical" nomenclature would be employed even though some units had not been named as yet, ordered the construction of standards, and fixed the date of July 1, 1794, as the deadline for national obligatory use of the metric system.[29]

Based upon this decree, the earliest "complete" system of metric weights, measures, and coins is presented in Table 4.3. The scientific work leading to the establishment of the natural standards was not yet finished, and there would be considerable problems to be faced on a number of other fronts. But, for the moment, France had done more in the space of a few years to bring about meaningful reform than other governments had accomplished over long spans of time.

Table 4.3

Decree of August 1, 1793

The First Metric System

Measures	Description
LINEAR	
10,000,000	quadrant of the meridian[a] = 5,132,430 fathoms
1,000,000	undetermined = 513,243 fathoms
100,000	grade[b] = 51,324 fathoms
10,000	undetermined = 5,132 fathoms
1,000	milliare = 513 fathoms
100	undetermined = 307 ft, 11 in, 4 lines
10	undetermined = 30 ft, 9 in, 6.4 lines
1	meter = 3 ft, 11.44 lines
0.10	decimeter = 3 in, 8.344 lines
0.100	centimeter = 4.434 lines
0.1000	millimeter = 0.443 lines
AREA	
square: 100 X 100 m	are = 10,000 sq m or 94,831 sq ft
rectangle: 100 X 10 m	deciare = 1,000 sq m or 9,483.1 sq ft

Table 4.3 (continued)

Measures	Description
square: 10 X 10 m	centiare = 100 sq m or 948.31 sq ft
CAPACITY	
1,000 cu m	cade = 1,051.33 pt or 78.9 bu
100 cu m	decicade = 105.14 pt or 7.89 bu
10 cu m	centicade = 10.5 pt or 0.789 bu
1 cu dm	pinte = 1.05 pt or 0.0789 bu
WEIGHT	
1,000 cu m of pure water	bar or millier = 2,044.4 lb
100 cu m "	decibar = 204.44 lb
10 cu m "	centibar = 20.444 lb
1 cu dm "	grave = 2 lbs, 5 gros, 49 gr
0.10 cu dm "	decigrave = 3 oz, 2 gros, 12.1 gr
0.100 cu dm "	centigrave = 2 gros, 44.41 gr
0.1000 cu dm "	gravet = 18.841 gr
0.10000 cu dm "	decigravet = 1.8841 gr
0.100000 cu dm "	centigravet = 0.18841 gr
MONEY	
0.100 grave of silver	franc = 188.41 gr

[a]In the decree the actual formula used for the values was: quarter of

Table 4.3 (continued)

---

the meridian = 5,132,430 fathoms = 30,794,580 feet = 369,534,960 inches = 4,434,419,520 lines; 1/10,000,000 part = 3 pieds, 11.44 lines. See Métrologies constitutionnelle et primitive, comparées, entre elles et avec la métrologie d'ordonnances (Paris, 1801), vol. 2, pp. 11-12.

[b]The grade, a decimal degree of the meridian, was actually the oldest unit of the metric system, antedating the meter, since it was employed in some pre-metric geodetic experiments.

The decree of August 1, 1793, called attention to the importance of establishing this system throughout France by the summer of 1794, but the violence and turbulence caused by the regime made implementation an arduous process to say the least. First of all, the government suppressed the Academy and purged all undesirable or suspect members on August 8, and this not only caused upheavals in the personnel entrusted with metric system development, but altered most of the existing procedures. The Academy was replaced by the newly constituted National Institute of Sciences and Arts, some of whose members were simply political lackeys devoid of qualifications, and it tried to continue the delegated scientific responsibilities. But with so many new members, the pace of metric refinement and acceptance by the general populace was slowed appreciably.

To complicate matters, but in keeping with the growing radicalism of the present Jacobin leadership under Danton, the calendar itself was altered abruptly to conform to the general decimal mania. By a decree of November 24, 1793, the new year was reckoned from the establishment of the French Republic, fixed at September 22, 1792, the day of the autumal equinox. The year was divided into twelve months as before, but each of these months was divided into 3 weeks, or decads, of ten days each. This was in conformity with the soon to be rejected program of dividing each day into ten hours and the circle into 400 degrees. The National Convention named a committee, chaired by G.C. Romme (1750-95), to create a new calendar and even appointed Fabre d´Eglantine (1750-94), a poet-dramatist, to rename the months of the new revolutionary year. Without going into any detail on a scheme that was ill-conceived, and that stood as little chance of success as the clock and circle changes, the committee grouped the months according to the seasons (vintage, foggy, frosty, snowy, rainy, windy, bud, flower, meadow, reaping, heat, and fruit) with a different name for each season. Even though this calendar persisted until 1806 (more than a decade longer than the clock aberration), its eventual doom can be blamed largely on the government´s persistent de-Christianization efforts, the eradication of the Sabbath due to 10-day weeks, the addition of a festival season to make up for more than five lost days every year, and, perhaps most important as in the case of clock dials, the sudden changing of well-established patterns of life by legislative enactment. The government ignored the obvious fact that the Gregorian calendar was used almost universally and never proved inconvenient to scientists or the general public. Change merely for the sake of change failed. There was a profound reason to alter the metrology of the ancient regime. There was absolutely no reason whatever to tamper with the calendar, the clock, and the circle.

While these schemes were circulating, the Convention charged its Committee of Public Instruction with the task of creating an agency to oversee the implementation of the new system, and on September 11, 1793, it formed the Temporary Commission of Weights and Measures and gave it the task of executing the decree of August 1, 1793. (This commission was later replaced in April, 1795 by the Temporary Agency of Weights and Measures, that was itself suppressed under the Directory in February, 1796.) This Temporary Commission consisted of twelve members: Borda, Brisson, Cassini, Coulomb, Delambre, Haüy, Lagrange, Laplace, Lavoisier, Méchain, Monge, and Vandermonde. Meusnier and Tillet were dead, while Condorcet was under arrest by the Revolutionary government. As president of this commission, Borda issued an elaborate report and sent it to the Committee of Public Instruction. However, Delambre and Méchain continued to measure the triangles as actively as the chaotic revolutionary conditions in France warranted. In addition to being arrested and deprived of ordinary facilities to carry on their work, they met with little sympathy and cooperation on the part of officials and the native population, and experienced great difficulty in erecting and maintaining their signals since they were often mistaken for spies. Such conditions threatened the entire enterprise.

The situation worsened on December 23, 1793, when the Committee of Public Safety dropped some of the members of the Temporary Commission for political reasons, these being Borda, Lavoisier, Laplace, Coulomb, Brisson, and Delambre. Philippe Buache (fl. 1760-1800), Jean Henri Hassenfratz (1755-1827), and Gaspard Clair Prony (1755-1839) were added. Hassenfratz soon thereafter replaced Lagrange as president, Haüy fulfilled the functions of secretary throughout the period, and Berthollet became treasurer succeeding Coulomb. This make-shift committee issued two progress reports on January 19, and March 5, 1794.

Amid these distressing events, the metric system was born. Of all the innovations of the Revolution, this was the most long-lasting.

5
COMPLETION OF THE FRENCH REVOLUTION: THE METRIC SYSTEM

The Law of April 7, 1795

The Temporary Commission of Weights and Measures, the National Institute of Sciences and Arts, and the government worked at a feverish pace after the decree of August 1, 1793, to finalize all aspects of their new metrological system. This entailed completing the series of primary units, providing appropriate multiples with Greek prefixes and submultiples with Latin prefixes (hitherto this bi-lingual division had not been proposed), and establishing a standard for mass. During this period from the creation of the first metric system to the law of April 7, 1795, still commonly called the Law of 18 Germinal An III, numerous political problems arose to impede the work of the commission. Delambre, for example, had to interrupt his geodetic measurements undertaken to establish the final value of the meter for more than a year, and the definitive meter that was expected to replace the provisional meter of 1793 within two years did not appear until 1799. As for the standard for mass, experiments to determine the weight of a known quantity of distilled water were made by Lavoisier and Haüy, but only their results obtained by January 4, 1793, remain since Lavoisier died soon thereafter at the hands of the Jacobins. His conclusions produced a value of 18,848.25 grains based upon the pile de Charlemagne, and the government used this figure to establish the provisional kilogram in 1795. Finally, in the pendulum experiments, the team assigned that task found the length of a simple pendulum that beat seconds at mean temperature at the Observatory of Paris equal to 50,999.6 parts, and they employed this figure to determine the length of the pendulum. After they compared this calibration with the Peruvian fathom, the length of the pendulum at this particular site was fixed at 440.5593 lines. However, since the meridian arc was well established by now as the basis of the metric system, this series of experiments had little value other than satisfying scientific curiosity.[1] (But in coming generations, this research would contribute significantly to horological refinements.)

The principal historical importance of this era was that the commission defined precisely all of the primary units of measurement, supplied additional multiple and submultiple prefixes, established the provisional standards for length and mass, and began the distribution of secondary standards to certain urban centers. Five major units emerged. The meter was the primary measure of length equal to the ten-millionth part of the terrestrial meridian between the North Pole and the equator. For superficial measurements there was the are designed specifically for arable and residential land, and it equaled a square of ten meters on each side. The stere, a volume measure for firewood, was a cubic meter. For liquid and dry capacity measures, the liter was a cube of one-tenth of a meter. Finally, the absolute weight of a volume of pure water equal to a cube of one-hundredth part of a meter at the temperature of melting ice was the gram. (In time all of these definitions would be changed, just as the units in the first metric system would be altered significantly by the actions of this commission.) The multiples and submultiples of these five primary units as defined by the law of 1795 are presented in Table 5.1.

Table 5.1

Law of April 7, 1795

Multiples and Submultiples

Multiples		Submultiples	
PREFIXES		PREFIXES	
myria	= 10,000	milli	= 0.1000
kilo	= 1,000	centi	= 0.100
hecto	= 100	deci	= 0.10
deca	= 10		
UNITS		UNITS	
myriameter	= 10,000 m	millimeter	= 0.1000 m
myriagram	= 10,000 g	milliliter	= 0.1000 l
myriare	= 10,000 a	milligram	= 0.1000 g
kilometer	= 1,000 m	centimeter	= 0.100 m
kiloliter	= 1,000 l	centiliter	= 0.100 l
kilogram	= 1,000 g	centigram	= 0.100 g
hectometer	= 100 m	centiare	= 0.100 a
hectoliter	= 100 l	centime	= 0.100 franc
hectogram	= 100 g	decimeter	= 0.10 m
hectare	= 100 a	deciliter	= 0.10 l
decameter	= 10 m	decigram	= 0.10 g

Table 5.1 (continued)

Multiples			Submultiples	
decaliter	=	10 l	decistere	= 0.10 stere
decagram	=	10 g	decime	= 0.10 franc
decastere	=	10 steres		

Even though some of these units soon would be discarded (the myria-group in particular), the rationales behind these selections by the Temporary Commission were many and varied. The members envisioned the meter serving principally for the aulnage of cloth, while a half and double meter had many uses on land and sea. The decameter of approximately 30 old feet was ideal as an acreage chain, and the kilometer made a useful length for expressing itinerary distances and for boundary markers on roads. The centiliter, the smallest practical capacity measure, represented a small glass for brandy, liqueurs, and certain medicinal potions. The decaliter was approximately the equivalent of an ordinary goblet; its half and double were convenient for most liquids. The liter differed little from the old litron or the Parisian pint and, employed for both liquids and dry products, its half and double fulfilled many functions. The decaliter and double decaliter replaced the grain bushel; the half decaliter replaced the picotin. For large grain, salt, plaster, lime, and coal measures, there was the hectoliter. Or it and its double made an appropriate wine measure, while a half hectoliter served for ordinary grain sales. The kiloliter was nearly an old sea tun and was ideal as a measure of account. The are was the basic standard for expensive city land, gardens, and small estates; the hectare for most agricultural land. The myriare was large enough to measure communes, districts, and other political or territorial jurisdictions. Finally, the stere replaced the Parisian demi-voie of wood.[2]

To help assure the adoption of these new units, and to aid citizens in conceptualizing their relative values, the government published the following conversion ratios several months after the April law.[3]

Table 5.2

Comparison of Old and New Units

Metric Unit	Value in Old System
meter	0.841712 Parisian ells or 3.07946 ft
square meter	9.48306 sq ft
are	2.92687 sq perches of 18 ft each
cubic meter	29.2027 cu ft
liter	1.05130 Parisian pt
decaliter	0.788473 Parisian bu
gram	18.841 gr
decagram	0.327101 oz
kilogram	2.04438 lb

Conversion manuals, however, did little to curb popular resentment and rampant non-compliance. Some disenchanted individuals and groups, notwithstanding the rancor of the aristocracy and other special interest groups who saw their traditional metrological privileges slipping away, believed that the implementation and reverification procedures were too expensive, that the effort exerted by the government was a waste of time and money that could be put to use solving the critical economic problems and shortages, and that metrics caused too great an upheaval in peoples´ lives. They argued that the metrological dilemma could be solved simply by streamlining the existing Parisian system and then mandating its use throughout the country. Some claimed that elitist scientists concocted metric weights and measures for their own use and never bothered to consult anyone else. Others rebelled against Greek and Latin prefixes, the difficulty in pronouncing and remembering them, and the fact that they were not French.[4] There were even complaints that the system lacked uniformity and organization due to contradictory and ambiguous legislation. (Similar arguments would be commonplace in Great Britain and the United States in the future.)

Popular resistance such as this caused the Temporary Commission to reduce appreciably the number of new units and to slow down the projected conversion timetable. Besides extending the deadline by three years, the government in September, 1797, agreed to recognize certain pre-metric units and to allow certain areas to proceed along a conversion path according to their own schedules.

The hue and cry of the citizenry, however, were not calmed by these tactics, and this forced the government to reexamine the methodical nomenclature intensely for several years, finally labeling it unacceptable in 1800. There were two principal reasons for this decision. First, the names were thought to be too long for expressing items that had frequent usage. Second, the composition of these names led to multiple usages. The government reintroduced simple names, similar to one of the plans proposed by the Academy in 1792, on the grounds that they were short and had meanings different enough to avoid confusion. The methodical system was a dead issue at least temporarily, but the fact remains that this was the fifth list of names either proposed or adopted in eight years. Frequent changes of this sort made acceptance of the system all the more difficult. Table 5.3 presents the units involved in the government resolution of November 4, 1800; the spellings given are as they appear in the state documents. Of course, changes would be forthcoming again once the metric system caught the popular imagination.

Table 5.3

Change in Nomenclature

Law of April 7, 1795	New Names and Values
LENGTH	
myriamètre	lieue of 10,000 m
kilomètre	mille of 1,000 m

Table 5.3 (continued)

Law of April 7, 1795	New Names and Values
hectomètre	no name given of 100 m
décamètre	perche of 10 m
mètre	mètre of 1 m
décimètre	palme of 0.1 m
centimètre	doigt of 0.01 m
millimètre	trait of 0.001 m
AREA	
hectare	arpent of 10,000 sq m
are	perche carrée of 100 sq m
centiare	mètre carré of 1 sq m
CAPACITY	
kilolitre	muid of 1 cu m
hectolitre	setier of 0.1 cu m
décalitre	boisseau or velte of 10 cu dm
litre	pinte of 1 cu dm
décilitre	verre of 0.1 cu dm
centilitre	no name given of 0.01 cu dm
VOLUME	
stère	stère of 1 cu m
décistère	solive of 0.1 cu m

Table 5.3 (continued)

Law of April 7, 1795	New Names and Values
WEIGHT	
no name given	millier of 1,000,000 g
no name given	quintal of 100,000 g
myriagramme	no name given of 10,000 g
kilogramme	livre of 1,000 g
hectogramme	once of 100 g
décagramme	gros of 10 g
gramme	denier of 1 g
décigramme	grain of 0.1 g
centigramme	no name given of 0.01 g
milligramme	no name given of 0.001 g
MONEY	
franc	franc of 5 g of silver
décime	sol of 0.5 g of silver
centime	denier of 0.05 g of silver

Scientific Results

The measurement of the meridian arc was the most important duty thrust on any metrological commission, since the final determination established the standard for every measurement division except mass, and this operation involved a vast amount of labor in field observations and in calculations. The task of measuring the French meridian was divided by Delambre and Méchain, the former assigned the northern portion between Dunkirk and Rodez, a distance of 380,000 fathoms, the latter given Rodez to Barcelona, a distance of 170,000 fathoms. The reason for this unequal division was that the northern part lay in a much more accessible terrain, while Méchain's measurements were in the mountainous regions of Spain. Also, the northern part had been measured twice previously. Both scientists encountered numerous difficulties owing to the political unrest, but by 1798 they completed their monumental work and returned to Paris where an international commission appointed by the Directory proceeded to examine their findings. The experiments of Delambre and Méchain were precision-controlled. They made 1800 observations each of the altitude of the pole star, for example, to determine astronomically the exact position of the Pantheon of Paris that formed a point on one of their triangles. The result showed an accordance of one-sixth of a second of arc, a remarkable finding by late eighteenth-century standards. In measuring the base lines the largest distance measured in one day was 360 meters, and they achieved this by placing one rule before the other over that distance ninety consecutive times. In this extremely careful manner they measured base lines of 6000 fathoms (approximately 7 1/3 miles) long.

Consisting of delegates from the Batavian Republic, the Cis-Alpine Republic, Denmark, Spain, Switzerland, the Ligurian Republic, Sardinia, the Roman and Tuscan Republics, in addition to a French delegation, the international commission divided itself into three groups to examine separate aspects of the work accomplished, and they made additional calculations and verifications to establish absolute accuracy and reliability of the findings.

The first group of scientists made a comparison of the bar used in measuring the length of the two bases at Melun and Perpignan. They found that it corresponded exactly with the Peruvian fathom. Examining the fathom of Mairan, constructed from the length of a pendulum beating seconds at Paris, they found it to be 0.03413 line shorter than the Peruvian fathom.

The second group studied the meridian arc, and the actual length of the meter, by remeasuring the bases, reexamining the angles of each triangle, and finally computing separately their dimensions, employing different tables of logarithms. The report prepared by Van Swinden of the Batavian Republic found that the difference between the computed and measured lengths of the base at Perpignan was 0.160 fathom. Since the length of the Perpignan base was 6,006.25 fathoms (or semi-modules as they referred to them at this point in time), and that of Melun was 6,075.9 fathoms, together with the fact that they were 550,000 fathoms apart, the accuracy of these measurements was astonishing.[5] Since the value of one base computed from the value of the other base by the triangulation measurements did not check exactly, a problem arose as to how the calculations should be made. Should the group compute the work wholly from one base, use a mean derivation, or calculate part of the work from one base and the remainder from the other? This last possibility served as the preferred method; the calculations from Dunkirk to Evaux depending on the Melun base, and the southern part from Evaux to Montjouy depending on the Perpignan base.[6]

They computed the flattening of the earth at the same time, comparing their measurements with those made in Peru. The most important result was the calculation of the length of the quadrant of the earth's meridian at 5,130,740 fathoms (32,808,992 English feet), which gave 3 feet, 11.296 lines as the exact length of the meter instead of 3 feet, 11.442 lines, the length of the provisional meter promulgated by the law of 1793.[7]

The last group issued a report that determined the exact unit of mass, and two of its members constructed a standard kilogram according to the earlier specifications of Lavoisier and Haüy. The preparation of this standard required elaborate experimental work, but the scientists finally determined that the weight of a cubic decimeter of distilled water at its maximum density weighed in a vacuum was 18,827.15 grains (2.0421 pounds), the mean of the sum of the pile de Charlemagne.

From these group reports, Van Swinden prepared a general report and presented it to the government. Lenoir constructed the actual meter standard and compared it carefully with the existing fathom standards. This platinum meter became the definitive meter, and since it was deposited in the Archives of State, it was known thereafter as the Meter of the Archives. Two other platinum standards were made called the Meter of the Conservatory and the Meter of the Observatory. Iron standards or secondary standards followed and the government alloted them to the delegates. The last action of the scientists was to present a platinum kilogram to the government on June 22, 1799. After certification, these definitive standards became the basis of the metric system on December 10, and all prior standards were abolished.

Reaction and Resolution

With the scientific determination of the units and the construction of the primary standards finalized, all that remained was to establish the general adoption of the system. But conditions delayed this once again, and a mounting opposition developed. First of all, massive political upheavals worked against acceptance of these weights and measures. When it began on September 25, 1792, a republic committed to metric adoption was the ruling governing body of France, with the radical National Convention guiding the transformation until 1795. In 1795 the Directory replaced the Convention, and carried the metric system to completion until it fell violently in 1799. For one year thereafter the government was controlled by the Provisional Consulate, and then by the Consulate from 1800-04. In 1805 the Empire of Napoleon Bonaparte began and, until his abdication on June 4, 1814, the destiny of France was in the hands of a man who hated republicanism and all that it represented. (This, and later administrations, were an ominous portent for metric acceptance.) To make matters worse, the Bourbons reestablished the monarchy of the ancient regime in 1814 when Louis XVIII became king, followed by Charles X in 1824, who ruled until a revolution deposed him and the Bourbon House in favor of the Capetian-Orleans House of Louis-Philippe (r. 1830-48) in 1830. It was natural to expect these anti-reformist leaders to show little or no interest in reforms accomplished by earlier republican regimes. Constant political chaos, frequent political changeovers, and occasional political fear of popular reaction served to stifle national adoption of the metric system.

Furthermore, there was a serious lack of secondary standards constructed and distributed at state expense. How were many communities to know what the standards were when they never received any copies? Also, the Temporary Commission was abolished much too soon, thus eliminating the new system's overseers. The situation became even more desperate when no attempt was made to overhaul the metrological officer corps. There were still hundreds of state appointed officials responsible for inspection, verification, and enforcement procedures, as well as thousands of private individuals who used their privileged status for private gain. Abuses and frauds flourished since non-conformity to the dictates of the new laws brought substantial monetary kickbacks to these officials. Just as in Great Britain, bureaucratic ineptitude and dishonesty knew no national boundaries.

These conditions brought about the law, or consular decree, of November 4, 1800. Even though decimal calibration remained intact in order to gain eventual popular acceptance of the metric system, the government reintroduced the old French names, save for the meter, stere, and solive, and applied them to metric units. A decree of September 22, 1801, even provided that the old names be inscribed on the metric weights already constructed, and that either one system or the other could be employed. Simply put, the government panicked and sacrificed metrological conformity to popular dissent and confusion. For twelve years France had two systems of weights and measures operating side by side. The metric system dominated government business, while this makeshift system was used in the provinces.[8]

While these new laws tended to weaken the integrity of the metric system, they at least preserved its fundamental feature of decimal division. But continuous travails, popular resentment, and government bungling during the first decade of the nineteenth century, led to a decree of Napoleon that threatened the very existence of the metric system. Despite the objections of Laplace and other distinguished scientists, Napoleon issued a decree on February 12, 1812, that established a system of weights and measures called the "système usuelle." In this remodeled system, metric weights and measures were altered in size in order to bring them into harmony with the units of the ancient regime familiar to commercial, business, and other interests. The legal metric system had to be taught in the schools, and its use was obligatory in all official government transactions.

The usual system defined its terms in those of the metric system and included some units popular to various sectors of society. For example, the fathom became 2 meters, divided into 6 feet, each foot of 0.33 meter. The foot, in turn, became 12 inches, each inch consisting of 12 lines. For the measurement of cloth there was the ell, equal to 12 decimeters and divided into many fractional subdivisions. These divisions were marked along one face of any metric scale or measure, while the opposite side retained the official metric divisions. The government provided other pre-metric units for retail business. For grain they allowed the boisseau or bushel, defined as 1/8 of a hectoliter, together with a double, half, and quarter bushel. Even the liter was divided into halves, quarters, and eighths. The pound became 500 grams or a half kilogram, with subdivisions of 16 ounces of 8 gros each. In tabular form the principal weights and measures included in these drastic changes were as follows.

Table 5.4

Principal Weights and Measures of the Systѐme Usuelle

Measures	Description	Values
LENGTH		
toise	none given	2 m
aune	consists of 1/2, 1/3, 1/4, 1/6, 1/8, 1/12, and 1/16 subdivisions	1.20 m
pied	1/16 toise	333 mm
pouce	1/12 pied	27.75 mm
ligne	1/12 pouce	2.33 mm
CAPACITY		

Table 5.4 (continued)

Measures	Description	Values
double boisseau	none given	0.25 hl
boisseau	"	0.125 hl
demi-boisseau	"	0.0625 hl
quart de boisseau	"	0.03125 hl
litre	consists of 1/2, 1/4, and 1/8 subdivisions	1 l
none given	subdivisions of 1/2, 1/4, 1/8, and 1/16 liters for wine and other liquids	none given
WEIGHT		
livre	16 ounces	500 g
demi-livre	8 "	250 g
quarteron	4 "	125 g
demi-quart	2 "	62.5 g
once	8 gros	31.25 g
demi-once	4 "	15.625 g
quart d´once	2 "	7.8125 g
gros	72 grains	3.90125 g

This awkward system remained in force for almost a quarter century, and it resulted in further confusion and postponement of metric system adoption. By 1837 some elements of the public, together with a number of government ministers, began a campaign to combat the major causes of resistance to metrication, that over the years became more and more entrenched. The usual system (commonly labeled the customary or traditional system in other countries or cultures) had developed a tradition of its own by now, and it satisfied the francophiles who defended its French nomenclature. Metrics to them represented the hated Revolution and its concomitant anti-clerical policy. Its radicalism was the work of government-sponsored scientists, and they believed that it helped contribute to the general and social tremors.

To overcome these serious obstacles, metric defenders stressed the total uniformity and cohesiveness of the system, its simplicity, and the immediate and long-range benefits to international commerce. They lauded its decimal structure that simplified calculations and eradicated the cumbersome employment of fractions. The basis of the system was the most scientifically sophisticated and accurate ever achieved. It had a reasonable number of denominations, with multiple and submultiple prefixes that were easy to remember. As an aid to mathematics, it was without parallel. Its standards were easily duplicated, it was time-tested, and it served the needs of all sectors of French and world society. (All of these arguments would be seriously debated by the British and other European peoples during the ensuing years.)

These pronouncements convinced the monarchy after a vigorous series of debates in several committees, for out of their sessions came amendatory legislation. The commissioners were adamant in refusing to accept any alteration of the system worked out before the Napoleonic era by so many dedicated scientists and government personnel. They saw the system enacted into law in 1812 to be a serious mistake. On July 4, 1837, their arguments won out, and an act was passed by the Chamber of Peers and the Chamber of Deputies that repealed the usual system, and decreed that after January 1, 1840, the only legal system of weights and measures allowable for use throughout France was the decimal metric system. Stiff penalties for non-compliance awaited those who ignored this law. The metric system was on its way to becoming a permanent French contribution to metrological and scientific accuracy. Its rapid adoption by other nations in the decades that followed is ample testimony to its arduous struggle during these tumultuous years.

6
BRITISH RESPONSE: THE IMPERIAL SYSTEM

While France attempted to free itself from the quagmire of the usual or customary system during the 1820s, and to establish the metric system as the only legal basis of weights and measures, England labored to put the finishing touches on its own metrological overhaul known as the imperial system. Not nearly as revolutionary in its unit simplicity or in its scientific bases for physical standards as that established officially by the French in 1837, it was, nevertheless, the most advanced and sophisticated metrology ever created in the British Isles. Compared to what existed over the previous thousand years, it was a ray of light out of a metrological tunnel darkened by unit profusion, imprecise standards, and bureaucratic failure. Its future was limited due to the impending struggle with metrics, and the latter's eventual victory, but during the remainder of the nineteenth and the early decades of the twentieth centuries, it held its own and served well the needs of British society, especially its commercial and technological sectors.

When George IV (r. 1820-37) became king of Great Britain in 1820, the British weights and measures law was more than eight centuries old. In this period that began with a decree of King Edgar (r. 959-75) in the tenth century calling for just weights and measures, various kings, ministers of state, concerned citizens, and legislative and bureaucratic personnel strove to bring order, uniformity, and precision to what appeared, at times, to be a hopeless morass of conflicting systems and personal interests. The principal emphasis of the laws before the nineteenth century was on unit standardization: 64 weights and measures were defined and accepted as national units. These units were a small fraction indeed (far less than 1/1000 percent) of the total number of weights and measures commonly employed throughout the British Isles, but they were the only ones ever given a legal status by the central government, or at least they were the only ones ever mentioned in any form of decree, order, ordinance, or statute that constituted official London endorsement.[1] Beginning with the commissions of the early nineteenth century, however, the emphasis of metrological law shifted drastically to the establishment of physical standards that were of high quality, craftsmanship, and reliability; to the appointment of supervisory personnel who would bring consistency and honesty to the regulation of weights and measures; and to the creation of quality control systems to protect citizens against frauds. This is not to say that unit standardization assumed a lower status in the law, but only that nineteenth-century legislators directed their energies to providing methods for maintaining and enforcing acceptable units and physical standards, rather than in describing or defining their physical dimensions. The law of 1824, together with one in 1878, helped to settle this millennium-old unit problem of British metrology until major amendatory legislation appeared in 1963.[2]

Legislation After 1824

The Imperial Weights and Measures Act of 1824, together with legislation issued in a fifty-year period thereafter, revolutionized to a considerable degree certain aspects of British metrology.[3] All previous decrees, assizes, orders-in-council, ordinances, and statutes were repealed. Eight hundred years of metrological law were erased overnight. Parliament broke irretrievably with the past and created a new law, one that determined the fundamental course of British weights and measures until the seventh decade of the twentieth century. These steps led to a drastic reduction in the number of legal measurement units; those retained acquired new and extensive applications. In place of those traditional tasks performed by thousands of different weights and measures officials, a new inspectorate secured all metrological duties. Quality controls took on a new importance in the law with special emphasis on consumer protection against frauds and irregularities in retail and wholesale merchandizing; in weighing and measuring procedures; and in commercial sales. This legislation was the most substantial achievement in British metrological history before the twentieth century.

There were only three weights and measures established as imperial standards under the act of 1824; all other weights and measures were based on these alone. This was a radical departure from earlier practice.

STANDARD	DESCRIPTION IN THE ACT
yard	The straight line or distance between the centers of the two points in the gold studs on the straight brass rod in the custody of the Clerk of the House of Commons engraved "Standard yard, 1760;" the temperature of

	the rod when verified was 62° F.
troy pound	The standard brass one pound troy weight made in 1758 and retained by the Clerk of the House of Commons.
gallon	The standard measure of capacity for dry products and liquids having a volume of 10 avoirdupois pounds of distilled water weighed in air against brass weights with the water and air at 62° F. and the barometer at 30 inches of mercury.

The yard standard—Bird´s brass rod constructed in 1760—became the only "measure of extension" whereby linear, superficial, and solid measures were derived. All measures of length were reckoned now in parts or multiples of this yard. The standard, immured in a recess in one of the stone walls of Commons under the custory of the Clerk, could only be removed decennially for testing purposes. Parliament ruled that if this standard were lost or injured in any way, it could be restored by reference to the length of a pendulum vibrating seconds of mean time in the latitude of London in a vacuum at the level of the sea. The proportion between the physical standard and the "invariable natural standard" in inches was 36 to 39.1393.

The troy pound—Bird´s model of 1758—became the primary standard for all other weights. The House of Commons retained it along with the standard yard. One-twelfth of this pound (the latter consisting of 5760 grains) made an ounce; one-twelfth ounce made a pennyweight; and one-twenty-fourth pennyweight constituted one grain. Seven thousand troy grains made an avoirdupois pound; one-sixteenth of the latter was an avoirdupois ounce; and one-sixteenth of this ounce made the avoirdupois dram (see Figure 11). The troy pound could be reconstructed in the event of some injury by reference to the weight of a cubic inch of distilled water weighed in air by brass weights at a temperature of 62° F. and a barometric pressure reading of 30 inches. Parliament legalized the weight of this particular quantity of water at 252.724 grains.

The gallon was the only standard used to derive the other measures of capacity. All extant gallon standards—such as the wine gallon of 231 cubic inches and the ale gallon of 282 cubic inches—were now illegal. The actual physical standard was not constructed until after the passage of the act. It was made of brass and had a diameter equal to its depth. At the same time a bushel was constructed of gun-metal, with a diameter equal to twice its depth. Since all capacity measures were figured in parts or multiples of the gallon, the quart became the fourth part, and the pint, the eighth part of this standard. The peck was two gallons; the bushel, eight. The quarter (more commonly called a seam in earlier periods) consisted of eight bushels. Originally the gallon was 277.274 cubic inches, but a more accurate determination made during 1931-32 corrected this figure to 277.421 cubic inches (see Figures 12 and 13).

The act of 1824 determined the course of British metrology for the next half century. Amendatory legislation over the next five decades served simply to refine certain provisions in this pioneering enactment until the act of 1878 reconstructed many of its basic provisions, especially those concerning inspection, verification, and enforcement duties of weights and measures personnel. For instance, an act in 1835 authorized inspectors to compare and verify all weights and measures in their own districts. Probably resulting from the fact that local merchants were recalcitrant in relinquishing their old standards, this was the first time in English law that the responsibilities of inspection and verification were required to be executed by the same local officer. Great Britain was on the verge of creating one of the most efficient metrological officer corps in European history. Parliament also insisted that any contract that included weights or measures unauthorized by this and other acts were illegal. The only legal denominations of weights and measures permissible after 1835 were the following.

Table 6.1

Legal Imperial Weights and Measures after 1835

Length	Capacity	Weight
100 feet	bushel	Avoirdupois:
1 chain of 66 feet	half-bushel	56, 28, 14, 7, 4, 2, and 1
or 100 links	peck	pounds[a]
1 perch of 16 1/2 feet	gallon	8, 4, 2, and 1 ounces
10 feet	half-gallon	8, 4, 2, 1, and 1/2 drams

Table 6.1 (continued)

Length	Capacity	Weight
2 yards	quart	240, 120, 72, 48, and 24
5 feet	pint[a]	grains or 10, 5, 3, 2, and
4 feet	half-pint	1 pennyweights
1 yard	gill	Troy Bullion:
2 feet	half-gill	500, 400, 300, 200, 100, 50,
1 foot	quarter-gill	40, 30, 20, 10, 5, 4, 3, 2,
1 inch[b]	Drugs:	and 1 ounces
	4, 3, 2, and 1	0.5, 0.4, 0.3, 0.2, 0.1, 0.05,
	fluid ounces	0.04, 0.03, 0.02, 0.01, 0.005,
	4, 3, 2, and 1	0.004, 0.003, 0.002, and 0.001
	fluid drams	ounce
	30, 20, 10, 5,	Decimal Grain:
	4, 3, 2, and	4,000, 2,000, 1,000, 500, 300,
	1 minims	200, 100, 50, 30, 20, 10, 5,
		3, 2, and 1 grains; 0.5, 0.3,
		0.2, 0.1, 0.05, 0.03, 0.02,
		and 0.01 grain

[a]Exceptions were made for weights above 56 pounds, for wooden or wicker measures used for selling lime, and for glass or earthenware jugs or

Table 6.1 (continued)

---

drinking-cups employed in retail establishments.

[b]It was permissable to divide the inch into 12 duodecimal, 10 decimal, and 16 binary equal parts.

Other significant pieces of amendatory legislation were passed in 1847, 1855, and 1858. In the first, Parliament established rules for weighing goods and carts in markets, made market authorities provide weighing houses and scales for general use by merchants and customers, and ruled that market weights and measures had to conform to Crown standards. It is obvious that there was still considerable circumvention of the acts of 1824 and 1835. In the enactment of 1855 Parliament provided for the storing of four copies of each imperial standard yard and pound, known as the Parliamentary Copies and constructed in conformity to those described in the act of 1824, at the Royal Mint, the Royal Society, the Royal Observatory, and the New Palace at Westminster, commonly called the Houses of Parliament. The emphasis here is clear: it was imperative that common access to reliable standards be provided in case of disputes. The lack of such standards is amply documented in earlier discussions. Finally, Parliament in 1858 mandated that inspectors examine the weights and measures of anyone selling goods in streets and public places. Besides offering further proof of non-compliance, the law was now extended to every conceivable commercial transaction. Although further resistance to these dictates was inevitable, the web of the law designed to protect buyers from unscrupulous sellers expanded more in this one decade than it had over the previous thousand years.

Lastly, acts passed in 1864, 1865, and 1870 provided for the authenticity of local standards and recognized the giant strides made by the metric system since 1837. The first, the Metric Act, legalized the use of metric terms in contracts, but not the use of metric units in trade. This followed a decision of a Select Committee appointed in 1862 by Parliament that opted for a complete metric adoption, but not until the general public gave its sanction. Metrics made its first major inroad into British life; a modest one to be sure, but the stage was set for more serious challenges later on. The second act transferred to the Board of Trade the imperial standards described in the act of 1824, and gave the Board the principal responsibility for making periodic inspections of these standards every ten years. The last act made this Board responsible for the creation and custody of the standard weights for weighing and testing the coin of the realm.[5] For the first time in British metrological history a qualified agency took control of one entire aspect of weights and measures. This was a further portent for future developments of far-reaching implications.

However, in other areas of the British Isles outside of England, imperial metrology spread at a much slower pace, and acceptance of these laws met stiff resistance. (Naturally this was to be expected, since there was continuous refusal to obey the new laws in England itself.) For example, even though the Act of Union in 1707 dictated that English standards apply automatically to Scotland, it was not until 1835 that the use of Scottish weights and measures became an offense punishable by fines, and some old Scottish units such as the "Scot's acre," boll, forpit, tron pound, stoup, lippie, jug, choppin, mutchkin, ell, drop, and stone continued well beyond 1835. The same situation held true for Ireland, but the time interval from the first official adoption of English standards to the passage of the imperial legislation was much greater than in Scotland, since the Irish Act of 1495 established the principle that all English statutes concerning the "public good" were applicable in Ireland, and this included English weights and measures legislation. Furthermore, two Irish acts of 1695 and 1705 provided for the deposit in the Irish Exchequer of avoirdupois weights from 56 pounds to 1 ounce as the standards of Ireland, and for the appointments of weigh-masters to execute all duties necessary in enforcing compliance to these English standards. Just as in Scotland, the Irish continued to employ their traditional measures, especially the "Irish perch" and "Irish mile" until the end of the century, and in some counties long beyond this. Similar circumstances occurred in the Channel Islands, particularly in Jersey, where the "Jersey pound," "Jersey quart," pot, noggin, half-noggin, cabot, and sixtonnier enjoyed immense popularity. Disenchantment with the new laws here was enhanced further by the fact that the imperial system had to compete with the metric and old Norman systems. France was not alone in finding it difficult to establish metrological reform. People in the British Isles resisted change just as emphatically and for a longer period of time.

Victorian Standards

Parliament's serious concern over injury or loss to the primary imperial standards was almost prophetic. Bird's troy pound and yard-bar lasted only ten years. Both of them were destroyed when one of the major fires in British history gutted the Houses of Parliament in 1834.[6] Imperial metrology was now truly unique: it was the only major system in Europe with no primary standards.

Among the earliest steps taken to replace these prototypes was the appointment in 1838 by T. Spring Rice (Lord Monteagle), the Chancellor of the Exchequer under Queen Victoria (r. 1837-1901), of a preliminary commission to consider and report on the proper method of restoration. The members of this commission had served previously on other panels to consider the steps to be taken should the standards ever be impaired or lost. Consisting of George Biddell Airy (1802-92), the Astronomer Royal, who acted as chairman, and of Francis Baily (1774-1844), John Herschel (1792-1871), John William Lubbock (1803-65), George Peacock (1791-1858), J.S. Lefevre (1773-1856), John Elliot Bethune (1801-51), and Richard Sheepshanks (1794-1855), this distinguished group presented its findings to the government on December 21, 1841. Their report highlighted the fact that several conclusions in the pendulum experiments of 1824 had proved subsequently to be inconclusive, and that a repetition of them now would not guarantee the exact reproduction of the lost standard yard. It was mandatory, therefore, that those sections in the act of 1824 that provided

Figure 11. Imperial Avoirdupois Pound Weights of 1824.
(Crown Copyright. Science Museum, London)

Figure 12. Imperial Quart, Gallon, and Pint (1824) of George IV.
(Crown Copyright. Science Museum, London)

Figure 13. Imperial Bushel (1824) of George IV.
(Crown Copyright. Science Museum, London)

for the pendulum method of restoring the linear measures be repealed. Instead, the standard of length should be defined in future legislation by the length, or by the distance between two points or lines on any length, of a certain piece of metal or other durable substance, supported in a prescribed manner and at a prescribed temperature. (If a "natural standard" were the desired goal, this was a definite step backwards.) Also, the determination of the density of water, based upon English, French, Austrian, Swedish, and Russian experiments, that served as the basis for the standard pound, could not be calculated with a greater accuracy than 1/1200 part; whereas an accuracy a hundred or even a thousand times greater was not uncommon in scientific weighing operations. These natural constants, then, so essential to earlier nineteenth-century scientific committees, were abandoned (perhaps out of desperation and the emergency of the moment), and the commissioners declared that it would always be possible to effect the restoration of standards simply by using material copies that, previously, had been compared with them and found to be accurate duplications. They believed that the original standards could be restored without appreciable error if the government adopted this new method.

To accomplish these goals, the committee recommended that four sets of copies (once again called the Parliamentary Copies) of the standards of length and weight be made, in every respect similar to the legal standards and equivalent to them. One set should be enclosed in a hermetically sealed case embedded within the masonry of some public building. (This was the first mention in British history of providing air and humidity protection for any primary standards.) This special set would never be utilized without the sanction of an act of Parliament.[7] Finally, the superintendence of the construction of these standards must be entrusted to a scientific committee named by the queen.

The preliminary commission also considered the relationship between physical standards and measurement units on both the decimal and duodecimal scales. Eschewing decimalization, they recommended that no structural change be made in the values of the primary units of imperial weights and measures, or in the names by which they were commonly denoted; that the construction of the standards be given to eminently qualified personnel; that the parliamentary standard of length be one yard; and that the avoirdupois pound be substituted for the troy, since the avoirdupois was used everywhere, while the troy was generally unknown in many areas of the British Isles, thus making it superfluous. Though they did not favor the introduction of a decimal system of weights and measures, they spoke strongly for the early adoption by the government of a decimal coinage similar to that of the United States. (This would not become a reality until 1969.) Lastly, they recommended that no specific standard of capacity be adopted since the act of 1824 defined it as the weight of 10 pounds of distilled water. (Rarely has a panel of such eminent scientific spokesmen started with so much potential and ended with so little accomplishment.)

Victoria's ministers, acting upon this report, reappointed Airy to head a second panel—called the Standards Commission of 1843—that included the Marquis of Northampton, Lord John Wrottesley (1798-1867), Professor William Hallows Miller (1801-80), plus those who participated in the preliminary investigations. Baily performed the operations for restoring the standard of length until his death in 1844; Sheepshanks then inherited the task. Miller conducted the experiments relating to the standard of weight. Their joint report materialized in 1854.

The Philosophical Transactions for 1856 and 1857 testify to the rigors of eleven years of intensive experimentation. The committee members followed earlier recommendations: they substituted the avoirdupois for the troy pound and verified the newly constructed yard and pound standards with precision scales made by Sheepshanks, Baily, and Miller. The magnitude of these operations may be gleaned from the fact that, in the case of the linear standard, the number of micrometer readings for all the comparisons exceeded 200,000, and, among other things, it was necessary to construct an entirely new system of thermometers. Among some of the measures of length used to gauge the new standard were Shuckburgh's five foot brass scale of 1796, two iron standards made for the Ordnance Survey in 1826-27, a brass tubular scale of the Royal Astronomical Society, and a Kater scale made for the Royal Society.

For the pound standard, the committee examined a Kater brass troy Exchequer standard of 1834, three similar brass pounds stored in London, Edinburgh, and Dublin, a platinum and two brass troy pounds owned by a London scientist, and a Royal Society platinum troy pound.

Thousands of experiments followed, and hundreds of standards were built and melted down, until the committee gave its approval to a select few. The members tested the new imperial yard and pound exhaustively, and Parliament legalized them in the act of 1855.[8] Complying with the recommendation of the preliminary committee, the government sent copies of them to the four sites mentioned above.

The new imperial standard yard was a solid gunmetal bar, 1 inch square in section and 38 inches long. Its metal composition, being a bronze alloy, was in the weight proportion of copper 16, tin 2.5, and zinc 1. Baily selected this particular ratio after many experiments in which he sought to find a material, that in a 1 inch square section bar, would be stiff without being brittle and, at the same time, would not have excessive alteration of length with changes of temperature. Near each end of the bar was a well-hole, 3/8 inch in diameter, with the two centers 36 inches apart. These well-holes were sunk to the mid-depth of the bar, and at the bottom of each hole was inserted a small gold stud (about 0.1 inch in diameter), with the defining line for each end of the bar finely engraved upon it. These defining lines took the form of two horizontal lines parallel with the length of the bar, crossed at right angles by three other lines at intervals of about 0.01 inch, of which the center line of each stud was the actual defining line. The other lines were inscribed to help the eye of the observer when using the bar for comparisons of length with micrometer microscopes. The entire bar rested horizontally on eight bronze rollers, carried by levers in such a way as to avoid flexure of the bar and to facilitate its free expansion and contraction from variations in temperature. In this way, the gold studs and their defining marks were protected from damage. This standard was correct at a temperature of 62° F. with the barometer standing at 30 inches of mercury (see Figure 14).[9] Since the initial verification in 1846, many other comparisons have been made. The conclusion reached concerning the accuracy of this standard by 1946 was that it had shortened only by approximately 2 parts in 1 million since 1900. This was, without a doubt, the best physical standard England had ever produced up to this point in time.

The primary standard of the imperial avoirdupois pound was made of platinum. In form it was a cylinder with slightly rounded edges top and bottom, 1.15 inches in diameter, and about 1.35 inches high. Just below the top was a groove all the way around the cylinder to take the ivory lifting fork by which it was moved in use. This standard was finely engraved on top—PS. 1844, 1 lb. In comparisons since 1846 the pound diminished progressively by 50 parts in 100 million between 1846 and 1883, and by a further 19 parts in 100 million between 1883 and 1933. Again, this is a remarkable figure compared to the manufacture of previous standards.

The imperial measures of capacity remained as in 1824, based on the gallon of 10 avoirdupois pounds of pure distilled water at a temperature of 62° F. and a barometric pressure of 30 inches of mercury.

There were no major changes in the imperial standards after the act of 1855. Their statutory wording had been modified somewhat; their application broadened considerably; but their general design and physical appearance were not altered.

In 1866 the Exchequer, that had controlled English standards since 1066, ceased being a separate office of government and was amalgamated with the Audit Office. A Standards Department of the Board of Trade was created immediately thereafter, and the standards were given to this new agency whose director, the Warden of the Standards, deposited them within a fireproof room. (The Great Fire was, perhaps, the chief inducement.) During this same year, an act reappointed the Standards Commission as a Royal Commission to consider and report on the condition of the standards and on the subject of weights and measures in general. Once again headed by Airy, this committee presented, between 1868 and 1871, five comprehensive reports that contained many important recommendations. They favored the substitution of metric for troy weight in the mint, its use in customs, and its general encouragement in all sectors of the British economy. Never, however, did they opt for compulsion since it was believed that factory owners and other industrialists could arrange such matters without legislative assistance. History, unfortunately, would prove otherwise. (Again, an Airy-chaired committee would prove short-sighted; not in technological contributions, but certainly in long-range planning and foresight.)

These recommendations were carried out by the Weights and Measures Consolidation Act of 1878. This important statute reaffirmed the existing standards established in 1855, but the number of unit denominations was reduced even more sharply, the troy pound was finally abolished, and all distinction between dry and liquid measure was rejected.[10] Parliamentary Copies of the imperial standards, described in the act of 1855, remained in the custory of their respective scientific and governmental depositories, and the Board of Trade, upon the request of Parliament, constructed a fifth set that was stored in the Standards Department. If at any time either of the imperial standards of length and weight was lost or damaged, the Board could attempt a restoration or simply adopt one of the Parliamentary Copies. If any of these Parliamentary Copies were adversely affected, the Board could restore them by reference either to the corresponding imperial standard or to one of the other surviving Parliamentary Copies. The Board of Trade standards—called the Secondary Standards—could be restored in the same manner.

Based on these standards, Parliament promulgated a list of the acceptable imperial weights and measures units, together with their new definitions. The latter contrasts sharply with past practice.

LINEAR AND SUPERFICIAL

inch = 1/12 foot

foot = 1/3 yard

rod, pole, or perch = 5 1/2 yards

chain = 22 yards

furlong = 220 yards

mile = 1760 yards

rood of land = 1210 square yards

acre = 160 square rods or 4840 square yards

WEIGHT

grain = 1/7000 pound

dram = 1/16 ounce

ounce = 1/16 pound

stone = 14 pounds

hundredweight = 8 stone

ton = 20 hundredweight

troy ounce = 480 grains

CAPACITY

pint = 1/8 gallon

quart = 1/4 gallon

peck = 2 gallons

bushel = 8 gallons

quarter = 8 bushels

chaldron = 36 bushels

Commercial transactions made in the United Kingdom were illegal if weights and measures other than these, or their multiples, were employed. (This was a severe setback for the metric system, but the Weights and Measures (Metric System) Act of 1897 declared this exclusion null and void and validated all contracts employing metric units.) Tolls and duties had to be charged or collected according to these weights and measures. All articles had to be sold by avoirdupois weight, except for gold, silver, articles made from gold and silver, lace, platinum, diamonds, and other precious metals or stones; in these cases only the troy ounce or its decimal parts could be employed.

The last significant development during the Victorian era in these units and physical standards occurred in 1889 when the International Meter, of which the British copy is an authorized replica, was established by the 1st General Conference of Weights and Measures, a world-wide organization set up by the Metric Convention of 1875 to propagate and perfect the metric system. The prototype is preserved at the International Bureau of Weights and Measures at Sèvres, France, and is made, like its national copies, of platinum-iridium in the proportion of 90-10. Its form is a bar of special X-section designed to give maximum rigidity for minimum use of material. It bears on the exposed neutral plane two parallel defining lines separated, at 0° C., by a distance declared to be the meter by the 1st General Conference. The definition of the meter adopted in 1889 was refined in some details by the 7th General Conference in 1927. A proviso was added stipulating that the bar could be used only under standard atmospheric conditions, while supported on two rollers at least 1 centimeter in diameter placed 571 millimeters apart symmetrically under the bar in the same horizontal plane.

The British copy is made from an alloy and is number 16 in a series of 30 bars that were distributed by lot to the member states of the Metric Convention. Being about 102 centimeters long with a cross-section of modified X-form, it bears at one end the markings "0° C & 20° C A. 16 SIP GENEVE 1956" and on the cross-section "1." At the other end it is marked

"B.16" and "2" on the cross-section. It has been recalibrated at the International Bureau several times since the original certification in 1889. In 1956 the defining lines were removed by repolishing the two facets in the neutral plane on which the lines were engraved. The measure was reruled then so that meter lengths at 0° C. and 20° C. were delineated. For the special purpose of optical interferometry, an additional line was ruled at a position approximately midway between the terminal lines on the bar. The copy was recertified after the reruling, the latter undertaken by the Société Genevoise d'Instruments de Physique on behalf of the International Bureau (see Figure 16).

The International Kilogram, the British copy (number 18) being an authorized replica, was established in 1889, and its prototype is made of the same platinum-iridium alloy as the meter bars. Constructed in the form of a solid cylinder, it has a height equal to its diameter (39 millimeters). When the metric system was established during the last decade of the eighteenth century, the kilogram was defined as equal to the mass of a cubic decimeter of water at the temperature of maximum density (4° C.), and the first permanent representation of the mass was the Kilogramme des Archives, a solid cylinder of platinum. Since the original definition was found later to be imprecise, it was superseded in 1889 by the current definition that declared that the kilogram is equal to the mass of the International Kilogram; the latter equal to the earlier standard within the limits imposed by the uncertainties of measurement at that date.

The British copy has been verified at the International Bureau several times since 1889. The five values of its mass obtained during the 1889-1960 period in relation to the prototype show an average deviation from the mean value of only 7 millionths of a gram, better than 1 part in 100 million.

The Modern Inspectorate

Contemporaneous with these significant achievements in unit simplification and the manufacture of sophisticated physical standards was the creation of the most proficient metrological officer corps in British history. In the first phase of the imperial system, or prior to the act of 1878, Parliament was interested principally in providing and maintaining the primary standards and in verifying copies of these for the use of local authorities. The major functions performed then by inspectors, called examiners before 1835, were securing agreement between these standards and examining weights and measures apparatus in factories and mercantile establishments. After 1878 Parliament moved in the direction of total regulation of weights and measures used in trade, and the powers of inspectors increased proportionally. By the end of the nineteenth century, these officials had gained a virtual monopoly over all aspects of weights and measures control; thus eliminating the thousands of non-qualified personnel who dominated these tasks since the Middle Ages.

Securing this monopoly, however, was a long and arduous process. Although the act of 1824 had established general uniformity among weights and measures, and had provided for material improvements in the standards by which they were to be regulated, it made no changes whatever in the

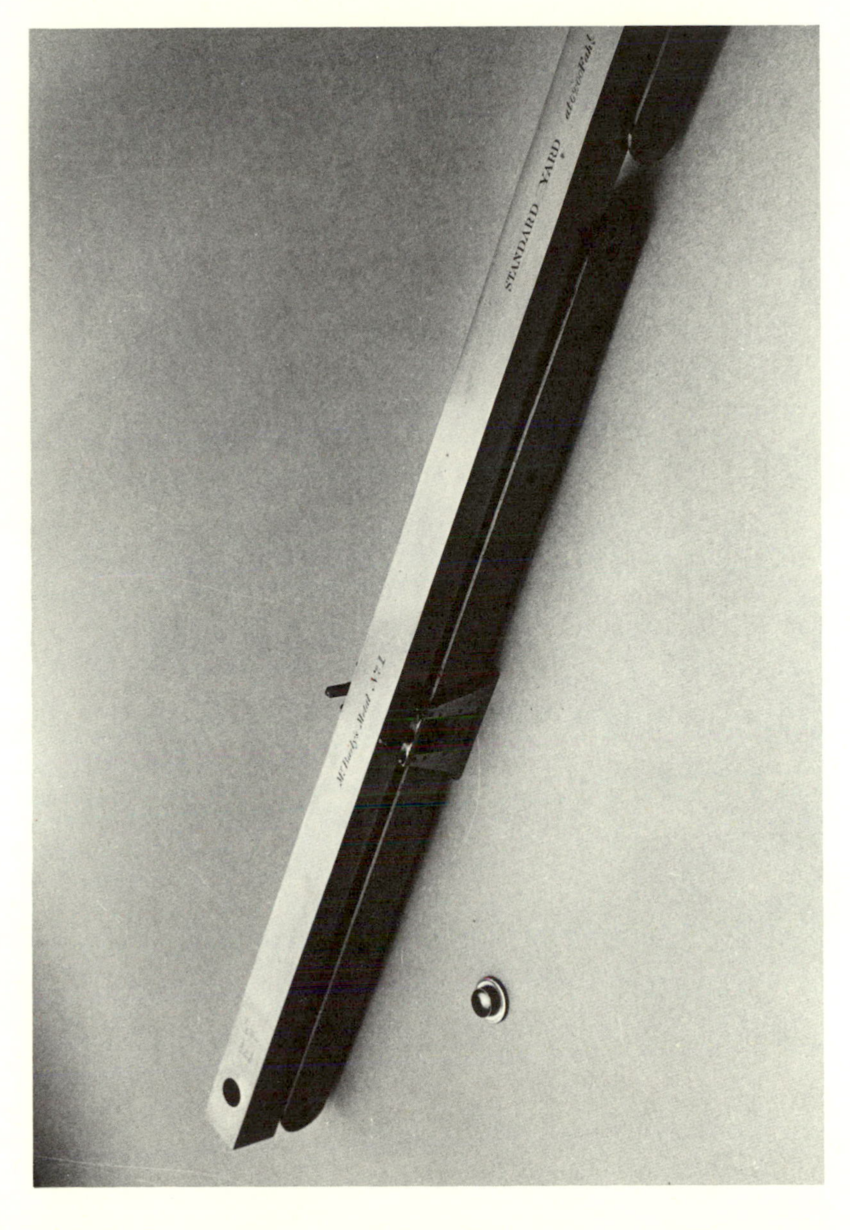

Figure 14. Imperial Yard of 1845. (Photo. John Moss).
(Furnished by the National Physical Laboratory, Teddington, Middlesex)

Figure 15. Imperial Apothecaries Measures of 1878.
(Crown Copyright. Science Museum, London)

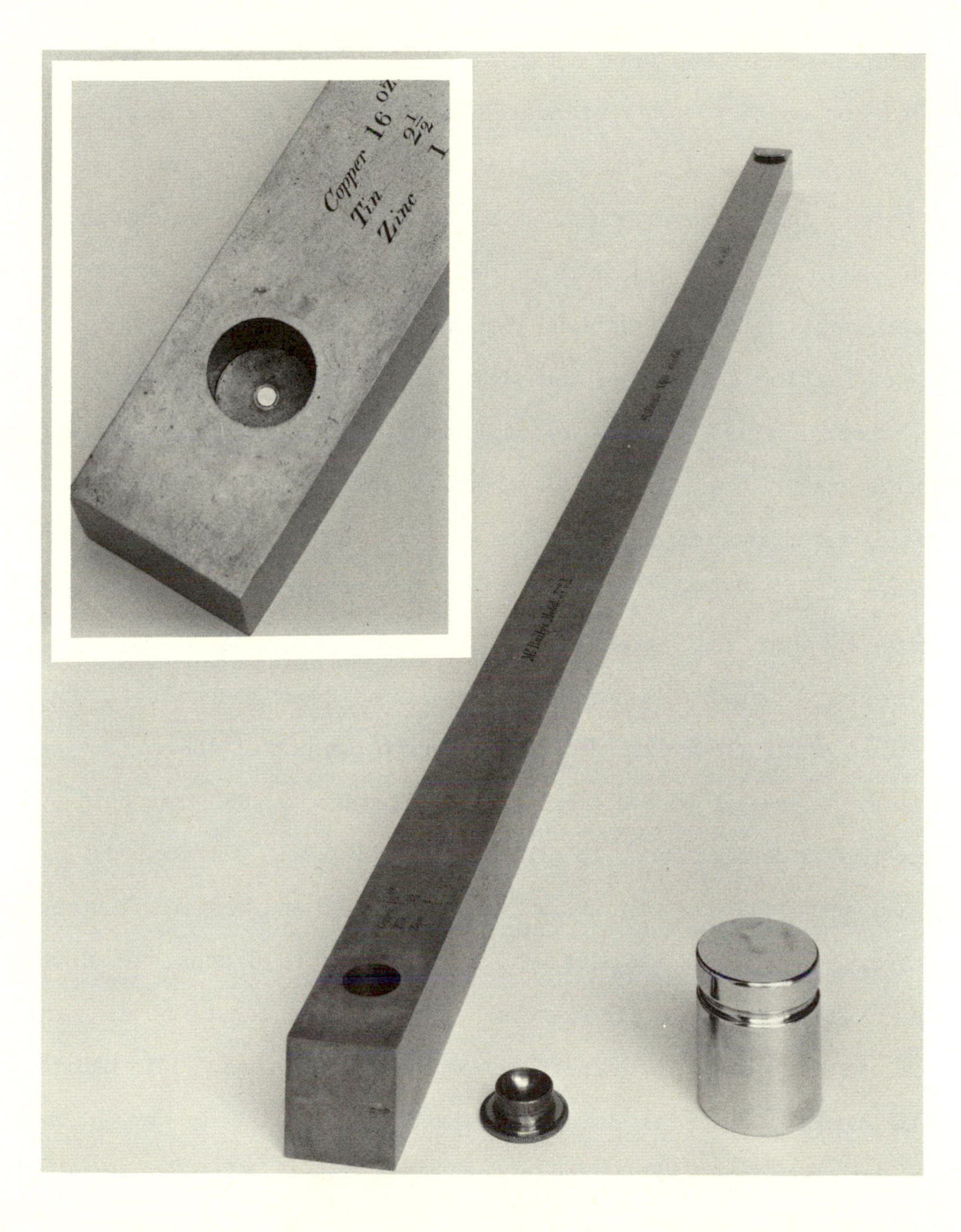

Figure 16. Imperial Yard and Pound kept at the Board of Trade.
(Crown Copyright. Science Museum, London)

system upon which they were administered. The act merely required that verified copies of the new standards be furnished by the magistracy of every county, city, and town, and that they be placed in the custody of such persons as the magistrates should appoint. In fact, no change in the existing administrative system is evident until 1835, when an act specified that both magistrates and justices must appoint inspectors and entrust them with examination (replacing the word "inspection" as of this date) and verification duties. Before accomplishing this, however, they were to determine exactly how many copies of the standards were needed in their respective territories and where the copies, verified and stamped, were to be permanently deposited. Only then could they appoint inspectors and fix the rates by which the latter were remunerated. It was not until a quarter of a century later that an act permitted town councils to appoint inspectors, a privilege extended to boroughs by an act of 1861.

Nevertheless, it became increasingly apparent by 1861 that this system had too many imperfections. To cite only two examples, although the act of 1835 had stipulated that the duties of examination and verification were to be performed by local inspectors, as late as 1870 other weights and measures officials were still exercising metrological functions in various places, particularly in certain metropolitan parishes. (Old habits are hard to break; if illegal units and standards persisted, why not excuse those who cherished past perquisites.) At these sites, local acts continued to give verification duties to local officials, and the only examination of weights and measures was carried out by annoyance juries.

In London, in contravention of Parliament's instructions, there was a stamper at the Guildhall who continued to verify weights and measures, while all examining was done by only two inspectors. Such practices can only lead one to assume that the system inaugurated in 1795, and strengthened several times thereafter, was interpreted locally as an addition to, and not a substitution for, the existing system inherited from medieval times. These and other problems warranted a solution; it would be achieved largely through legislation passed in 1878 and 1889.

The basic pattern for the administrative system of weights and measures was molded chiefly by the act of 1878. Parliament ordered local authorities in every county and borough to provide such denominations of standards as were necessary for the purpose of comparison with the weights and measures in each of their areas.[11] These authorities became:

AREA	LOCAL AUTHORITY
	In England
county	the justices in quarter sessions
county of London	the court of the lord mayor and aldermen
borough	the mayor, aldermen, and burgesses
	In Ireland
county	the grand jury in certain terms
borough	town council[12]

They alone fixed the places at which these standards were to be deposited and the methods to be followed in their stamping and verification. From time to time they appointed sufficient numbers of inspectors, and they charged them with the maintenance of the local standards. They suspended or dismissed inspectors, or appointed others, as circumstances dictated, assigning reasonable remuneration to each inspector for services rendered. They also delegated examination and verification duties to different inspectors.

In accordance with these and later instructions (most of which are still operative), an inspector, once appointed, entered into a "recognizance to the Crown." This personal performance bond acted as a guarantee for the proper performance of his duties; for the quick payment, at times stipulated by the local authority, of all fees received; for the safety of the local standards, stamps, and appliances for verification committed to his keeping; and for their immediate surrender, upon his removal or other cessation from office, to the person appointed to receive them. After the local official had designated when and where verifications must be performed, the inspector began his duties. With the local standards in his custody, he examined weights and measures. If he found them to be correct, he stamped them with a verification mark in such a way as to prevent frauds.[13] When appropriate, he affixed a name, number, or sign distinguishing the district that appointed him. He was required to register every verification in his own record book, and he was to give, if required, a certificate for the stamping.[14] He could enter any premises to verify and stamp weights and measures if the persons there were residents of his own district. If, however, he knowingly stamped a weight or measure of someone residing in a district where there already was an inspector legally appointed by another local authority, he was liable to a fine. Once it was properly stamped, a weight or measure was legal throughout the United Kingdom, unless found subsequently to be false.

Desiring still more control over their activities, Parliament empowered the Board of Trade in 1889 to appoint an officer to oversee the administration of weights and measures law within the jurisdiction of any local authority. The person selected had the right to do whatever was necessary to insure that metrological law was being carried out faithfully and dutifully.[15]

Many acts since 1889 have added to the powers of the inspectorate and to the means of controlling infractions committed by them. In that time Great Britain solved its inspection, verification, and enforcement dilemma. More than a millennium of trial and error came to an end.

7
BATTLE OF THE STANDARDS

British legislation during the nineteenth century created three outstanding metrological achievements. The number of official measuring units was reduced substantially until the government selected the most important among them by which a simple, integrated, and coordinated system of weights and measures could be applied to all aspects of British life. Physical standards reached a plateau unheard of in previous eras, especially from the standpoints of reliability, accuracy, precision, and durability. Finally, the newly formed inspectorate was a model for other nations to follow in establishing a permanent, professionally trained corps to perform the essential tasks necessary for the smooth functioning of metrological law. More positive accomplishments occurred in this single century than had ever taken place before.

Metrological reform plans, however, did not expire as a result of these nineteenth-century efforts. Quite the contrary, there were now additional plans advanced and their authors, with few exceptions, recommended programs that were much more revolutionary than those conceived before 1824. Later reformers rejected the imperial commissions for compromising on most of the critical issues. Influenced by the metric phenomenon, some argued that current legislation did not settle the problem of finding an acceptable standard. The new standards, they believed, were no better than the earlier ones since they were not based upon some invariable natural constant such as the French meter. Also, since no law introduced unit decimalization, or made significant alteration in the values and proportions of units, Parliament merely replaced one inferior system with another. Others remarked that no attempt was made to reform British coinage such as achieved by the United States in its decimalized dollar system introduced in the 1790s. As a result, new proposals sprang up that recommended substantial changes in the names, values, proportions, and denominations of units. Several "metric" or quasi-metric systems were advanced, in addition to more than one type of "universal" system—the latter offered in the hope that one could improve upon the French plan and eventually win world-wide acceptance. Because nineteenth-century government reform had only gone part of the way in altering the old metrology, the overwhelming emphasis in these later proposals was on radical change. Regretfully, government ministers dealt with them in exactly the same way as they had with those of the pre-imperial era. Government recalcitrance, however, merely postponed the inevitable, and this aided the cause of future metric adoption.

Post-Imperial Reform Proposals

Surprisingly, none of these radical proposals appeared until 1851. Whether metrologists needed that amount of time to assess adequately the successes and failures of the imperial system, or whether later amendments to the act of 1824 made it appear that substantial changes would be forthcoming, thus delaying serious opposition, or whether they were awaiting the final verdict on metrication efforts in France and elsewhere, one cannot say with certainty. Regardless of the actual cause, the first work to suggest an alternative to the imperial system was Henry Taylor's

The Decimal System, As Applied to the Coinage & Weights & Measures of Great Britain. Published in London in 1851, Taylor's (1784-1876) main purpose was to show how the imperial system could be recast into a decimal framework and, it was hoped, into closer alignment with the metric system. His ultimate purpose was to ease transition to a quasi-metric system in the expectation of eventual metric adoption throughout the British Empire. (Time would prove him correct.)

Taylor believed that the imperial commissions had done little to alleviate the unnecessary complexity among unit values and proportions. (A view shared by many of his contemporaries.) Even the post-1824 commissions, staffed with such brilliant scientists and public servants as Airy, Baily, Bethune, Gilbert, Herschel, Lefevre, Lubbock, Peacock, and Sheepshanks, had been negligent in this respect. Troy weight continued despite statutory injunctions limiting its employment. (Taylor refers here to the troy pound, rather than the ounce reserved officially for mint, gold, and silver purposes.) To make matters worse, the new avoirdupois weights had no intrinsic constituent parts since they were restructured according to a troy grain scale; further confusion resulted rather than less. A stone of 14 pounds was disadvantageous since it was either too large or too small for general use. Ten pounds was the perfect amount since it represented the exact imperial gallon of water. He saw no reason for retaining a hundredweight of 112 or 120 pounds, or tons based on these non-decimal aliquant figures; one hundred pounds was equally as convenient for expressing large quantities. Besides, binary subdivisions in government and commercial account keeping would be executed more easily by decimal notation, and the margin of error would be reduced sharply since fractions would be eliminated. (One of the most popular claims made repeatedly thereafter by metric spokesmen.) These long-standing deficiencies could be rectified quickly and easily if Parliament would decimalize all existing imperial units.[1]

In order to achieve a perfectly integrated decimal system, Taylor first subdivided the imperial pound of 7000 troy grains into 10,000 equal parts. Unless this were accepted beforehand, he argued that none of the subsequent decimal divisions and accretions would be possible. The change to a decimal scale would be felt in no way beyond the necessity of having a few small new weights added. Also, it was necessary to rename those submultiples constituting the pound since, in Taylor's words, the term "grain" no longer bore any reference to wheat or barley grains. In looking for a suitable name, he settled on the term "minim," because it denoted customarily the lowest unit of measurement in other systems, and in his new plan it was exactly that. The only other changes were the dropping of the name "pennyweight" in favor of "scruple," the reason being that the latter was shorter and, as a means of easing a future metric changeover, he favored the addition of the name "meter" to any weight describing a capacity measure; thus, 1 ounce-meter in lieu of the traditional 1 fluid ounce. (The retention of certain of these archaic names of pre-imperial vintage would damage his efforts seriously.) His decimalization scheme is outlined in Tables 7.1 and 7.2.

Table 7.1

Proportions Among the Weights in the Taylor Plan

minims	scruples	drams	ounces	pounds	stone	cwt	tons
1							
10	1						
100	10	1					
1000	100	10	1				
10,000	1000	100	10	1			
100,000	10,000	1000	100	10	1		
1,000,000	100,000	10,000	1000	100	10	1	
10,000,000	1,000,000	100,000	10,000	1000	100	10	1

Table 7.2

Dry and Liquid Capacity Measures in the Taylor Plan

Cubic Inches	Units
2.8	1 minim
28	1 scruple-meter (10 minims of water)
277	1 dram-meter (10 scruples of water)
2773	1 ounce-meter (10 drams of water)
27,727	1 pound-meter (10 ounces of water)

Table 7.2 (continued)

Cubic Inches	Units
277,274	1 gallon (10 pounds of water)
2,772,740	1 hundred-meter or firkin (10 gallons of water)
27,727,400	1 butt (10 firkins of water)

Besides the obvious advantages derived from decimalization, Taylor was enthusiastic about the fact that the new suggested ton of 1000 pounds represented exactly 16 cubic feet of spring water. He also pointed out that in bookkeeping all mention of quarters would cease, and reckoning would be done centesimally, in hundredweight and pounds. Since the total number of weights and measures was reduced significantly, and since their values and proportions were easy to remember, fraudulent practices would occur with less frequency. In order to cope with the expected resistance by local populations, he advised that measures such as quarts, quarters, pints, kilderkins, barrels, hogsheads, puncheons, kegs, ankers, runlets, tierces, pipes, and tuns, that persisted despite the reform of 1824, be retained strictly for sales purposes without relating them to any definite number of gallons.

Even though Taylor was a practical metrologist, his acceptance of traditional, non-imperial units that were forbidden after 1824 tarnished his basic proposals, and led to its eventual rejection. Also, his plan, like some of pre-imperial vintage, was incomplete, and this deficiency helped to insure its demise.

Following Taylor's conscientious effort of 1851, the most extensive decimal plan ever devised in the British Isles was published by the Cambridge University Press in 1855. In A Complete Decimal System of Money and Measures, William Henry Jessop, a mathematics professor at Trinity College, worked out a vastly simplified metrology that bore no relationship whatever to the imperial system, save for most of the names selected for the units. He decimalized all weights and measures in his system, and suggested an entirely new, and remarkably innovative, set of standards. Jessop's was the only massive reconstruction of imperial metrology during the nineteenth century, and it was far more radical in its approach and application than anything attempted by the pre-imperial reformers. He tried to do for English reformed metrology what the French had achieved for the metric system: a model for world-wide adoption. The French scheme won largely because it transcended national and ethnic boundaries. Metrics were truly universal; Jessop's plan, unfortunately, was inherently and intrinsically English.

Unlike Taylor, Jessop included linear and superficial measures in his proposal, thus rendering it much more immediately acceptable. The only incomplete facet dealt with volume measures, based largely on the fact that the pre-imperial and imperial systems did not have a sufficient number of units in this particular measurement division. Despite this flaw, this was a comprehensive system, not an unfinished or speculative ideal that required future experimentation and finalization.

For his primary standard of length, Jessop chose the digit, and defined it as one-sixty-fourth part of the length of a pendulum vibrating seconds in a vacuum at sea level at Greenwich latitude.[2] All other linear measures were formed by multiplying the digit successively by 10. Since the length of this pendulum was found to be 36.13929 inches, the digit was 1/64 of its length or 0.61155 inches. Based upon this standard, the other units, expressed in inches, were: link = 6.11551; pace = 61.15514; chain = 611.55140; furlong = 6115.51406; and mile = 61,155.14062.

In tabular form, the following decimal proportions existed among these units.

Table 7.3

Proportions Among the Linear Measures in the Jessop Plan

digits	links	paces	chains	furlongs	miles
1					
10	1				
100	10	1			
1000	100	10	1		
10,000	1000	100	10	1	
100,000	10,000	1000	100	10	1

When Jessop´s new units were compared with those in the imperial system, their values and proportions, expressed in imperial inches, in most instances, were remarkably similar.

Table 7.4

Comparison of Linear Measures in the Jessop and Imperial Systems

Unit	Jessop	Imperial	Proportion
digit	0.61	0.72	1:1.19
link	6.11	7.92	1:1.29
pace	61.15	60.0	1.02:1
chain	611.55	792.0	1:1.29
furlong	6115.51	7920.0	1:1.29
mile	61,155.14	63,360.0	1:1.03

Just as in the imperial system, 100 links equaled 1 chain, and 10 chains equaled 1 furlong, but the length of each of these three measures was diminished by one-third. The other linear measures were changed only slightly. The digit was diminished by less than a fifth and the pace was increased by one-fifteenth. Jessop purposely selected names that were in close proportional alignment with the values of his decimal units. The familiar inch, foot, and yard were eliminated altogether, since their respective values of one-twelfth, twelve, and three made them inappropriate for decimal nomenclature. It was upon the principles of this initial measurement division that Jessop built his entire metrological edifice.

Jessop made his new area measures the square of the new linear measures according to a strict decimalized structure. Since this division of measurement had to fulfill multiple usages, he felt that it was imperative to incorporate some well-known names from the old system to fill in gaps that were now evident, and to provide convenient measures to satisfy all of the needs of agriculture, mining, land surveying, municipal subdivisions, and the like. Thus, the perch, rod, and pole became strictly superficial measures to accommodate an insufficient number of linear measures, and he added the traditional acre and hide that had always designated square measurement. There were only slight differences between the old and new poles, square paces, and square miles, but the differences in the acres, square chains, square furlongs, and hides were considerable. The latter resulted from the fact that they now had to conform to a decimal arrangement irrespective of tradition and popular custom. Finally, the words "perch," "rod," and "pole" hitherto had been used to denote the same measure, rendering two of them superfluous. These names were even retained in the post-1824 imperial legislation, but they designated the same measure. In the new plan they signified three separate and distinct measures.

The new standard for superficial measures was the square digit, equal to 0.373 square inches. Based upon this unit, and multiplied successively by 10, the remaining area measures, expressed in square inches, became: perch = 3.739; square link = 37.399; rod = 373.995; square pace = 3739.951; pole = 37,399.512; square chain = 373,995.122; acre = 3,739,951.224; square furlong = 37,399,512.248; hide = 373,995,122.486; and the square mile = 3,739,951,224.863.

Compared with the corresponding imperial units, and rounded off to three decimal places, their values and proportions were the following.

Table 7.5

Comparison of Area Measures in the Jessop and Imperial Systems

Unit	New Sq In Value	Old Sq In Value	Proportion
square digit	0.373	0.529	1:1.416
perch	3.739	none	none
square link	37.399	62.726	1:1.677
rod	373.995	none	none
square pace	3739.951	3600.000	1.039:1
pole	37,399.512	39,204.000	1:1.048
square chain	373,995.122	627,264.000	1:1.677
acre	3,739,951.224	6,272,640.000	1:1.677
square furlong	37,399,512.248	62,726,400.000	1:1.677
hide	373,995,122.486	627,264,000.000	1:1.677
square mile	3,739,951,224.863	4,014,489,600.000	1:1.270

Even though this was a far more comprehensive listing than that available under imperial metrology, it could be argued strongly that making three separate measures out of the perch, rod, and pole only intensified popular confusion. These names were too ingrained in the popular imagination; they always indicated the same linear measure. Jessop would have been better advised to have eliminated all three.

The dram was the standard for mass. Jessop defined it as the weight of a cubic digit of distilled water at 62° F. and at 30 inches of barometric pressure, figures used by the imperial commissions in establishing the new imperial standards. Just as in linear and superficial measurement, all weights were formed by multiplying the dram successively by ten. In the avoirdupois scale, the cubic digit of distilled water (the new dram) totaled 2.111687753 imperial drams. He saw most of the existing imperial weights as useless, since there were great proportional differences among them, they were used only in certain industries or in certain places, and they were impossible to reduce to a decimal order due to their peculiar denominational values. Those retained in his system, reduced to three decimal places, and expressed in his new drams equal to 2.111 imperial drams, were the ounce = 21.116; pound = 211.168; stone = 2111.687; hundredweight = 21,116.877; manpower = 211,168.775; and horsepower = 2,111,687.753.

The old definition of horsepower—the weight that a horse can draw up a vertical height of one foot in one minute—was replaced by the weight that a horse can draw up a veritical height of one digit in one second, and Jessop assumed that this weight was equal to 10,000 of his new pounds. The relationships between the old and new systems of weight are as follows; horsepowers and manpowers are expressed in avoirdupois pounds, while all other weights are in avoirdupois drams.

Table 7.6

Comparison of Weights in the Jessop and Imperial Systems

Unit	New Value	Old Value	Proportion
dram	2.1	1	2.11:1
ounce	21.1	16	1.32:1
pound	211.1	256	1:1.21
stone	2111.6	2048	1.03:1
hundredweight	21,116.8	28,672	1:1.35
manpower	2522.2	2200	1.14:1
horsepower	25,222.7	22,000	1.14:1

Since a cubic inch of distilled water under the imperial system weighed 252.458 grains, a cubic digit in the new system weighed 57.74149 grains. The latter, Jessop's weight standard, can also be compared in the following manner.

Table 7.7

A Further Comparison of the Jessop and Imperial Weight Systems

Unit	New Grain Value	Old Grain Value	Proportion
dram	57.74	60	1:1.04

Table 7.7 (continued)

Unit	New Grain Value	Old Grain Value	Proportion
ounce	577.41	480	1.2:1
pound	5774.14	5760	1.0024:1
hundredweight	577,414.97	576,000	1.0024:1

For liquid and dry capacity measures the standard was the cochlear, the content of a cubic digit equal to 0.2287 cubic inches. As usual, each capacity measure was multiplied by ten to produce the next largest unit. Since one cubic digit (the cochlear) was equal to 0.228717243 cubic inches, the entire system, expressed in cubic inches, and carried to four decimal places, was cyath = 2.2871; pint = 22.8717; gallon = 228.7172; anker = 2287.1724; and butt = 22,871.7243.

Jessop chose the first two of these measures from ancient Greek metrology. There were three principal reasons for their inclusion. First, their relationship had been 10 to 1, hence they fitted perfectly into a decimal system. Second, their cubic capacities were very small. Third, they were used in Greek society for both liquid and dry products. (The inclusion of these ancient measures, coupled with their strange names to most British ears, would have the same effect as did the Latin and Greek prefixes on most French citizens.) Based on the old wine gallon of 231 cubic inches, the following table shows the comparative differences between the old and new systems.

Table 7.8

Comparison of Liquid Capacity Measures

in the Jessop and Imperial Systems

Unit	New Value	Old Value	Proportion
cochlear	0.22	0.27 (ancient Greek)	1:1.2
cyath	2.28	2.74 (ancient Greek)	1:1.2
pint	22.87	28.875	1:1.26
gallon	228.71	231	1:1.01
anker	2287.17	2310	1:1.01
butt	22,871.72	23,100	1:1.01

As shown above, there was only one significant alteration among any of these measures from those commonly employed in the imperial system. The utility and ease of application of this system were its distinguishing features. For example, the weight of a cubic digit of distilled water was a dram. Since a gallon contained 1000 cubic digits, it could be expressed in weight as 1000 drams, or 100 ounces, or 10 pounds, or 1 stone. As in the imperial system, the gallon held 10 pounds, but unlike the new gallon, the Parliamentary commissioners fixed the imperial standard arbitrarily.

The latter's size had been determined from the amount of water that it contained. Hence, the standard of weight had to be known before that of capacity could be found. In Jessop's plan, either of the two standards could be determined independently of the other. The gallon contained 1000 cocnlears, therefore it held 1000 drams of water. He selected the anker primarily because it was already in decimal order, being equal to 10 imperial gallons. (Again, this was an illegal imperial unit.) Lastly, he chose the butt instead of the pipe (both illegal imperial units) because the latter was primarily a wine vessel with limited universal product usage.

As stated before, the standard for dry capacity measures was the same as for liquids. In fact, the first four units in both capacity measurement listings were exactly equal in name and content. The larger measures for dry products, when compared with their imperial counterparts in cubic inches, appear in the following table.

Table 7.9

Comparison of Dry Capacity Measures

in the Jessop and Imperial Systems

Unit	New Value	Old Value	Proportion
gallon	228.71	277.274	1:1.21
bushel	2287.17	2218.192	1.03:1
quarter	22,871.72	22,523.909	1.01:1

Table 7.9 (continued)

Unit	New Value	Old Value	Proportion
last	228,717.24	225,239.096	1.01:1

The only appreciable difference occurs in the gallons. Jessop retained the name "quarter" because it was so well known, and because there was no measure of which it was now the fourth part.

Jessop´s only failure was to complete a scheme for decimalizing measures of volume; the only units provided were the cubic digit, cubic link, cubic pace, and floor (an awkward and poor choice for the largest measure). His basic problem in this particular measurement division was that there never existed any officially authorized volume measures in either pre-imperial or imperial metrology. Hence, whatever he created here would be totally foreign to British custom.

Jessop was unsuccessful ultimately in gaining either a serious government ministerial or parliamentary audience for his transformation of imperial metrology. Suffering the same fate as some of his predecessors, he was not a member of the government establishment, and his proposal was not initiated by an official directive. Aside from his inability to furnish a complete set of volume measures, his plan was the most comprehensive and simple of any ever proposed in the British Isles. It was decidedly superior to the imperial system in its decimalized structure and paucity of unit names, and would have represented a marked improvement. But is was definitely inferior to the metric system´s physical standards, universal nomenclature, and universal application. This was the last privately sponsored reform plan in British history. The future belonged to metrics.

World Metric Refinement

With the growing use of the metric system for scientific work throughout Europe, the precision of its fundamental units became more important than ever. Increased activity in geodesy brought about several new measurements of meridian arcs, and the new data made it possible to recompute the shape of the earth and the length of the quadrant. Any change in the latter affected the length of the meter as the primary unit of measurement. This problem was demonstrated by Friedrich W. Bessel (1784-1846) in 1844, and shortly thereafter by General Theodore Friedrich von Schubert (1789-1865) of the Russian Army, Colonel George Everest (1790-1866) of the British Army, and Captain A.R. Clarke (1824-1902) of the British Ordnance Survey, all of whom conducted experiments to determine more accurately the shape of the earth. Because of this work, some scientists now began to have grave reservations concerning the continued use of the quadrant of a great circle as the basis for the metric system, since it varied in different places and required an exact knowledge of the actual shape of the earth.

These questions were purely scientific of course, and did not influence the practical development of the system either in France or elsewhere, but they provoked vigorous discussion. With the world's expositions that began in London in 1851, an opportunity arose for all sectors of society to examine and appreciate the benefits of an international system of weights and measures, while statistical congresses saw advantages resulting from decimalization. Most important among the former was the Paris Exposition of 1867 that encouraged the adoption of the metric system throughout the world. (The metric system being of French creation and design, this was to be expected.) A committee formed there advocated its use in secondary schools, scientific publications, public statistical studies, postal services, customs regulations, and government documents.

During this same year the International Geodetic Association, composed of delegates from various European countries, met in Berlin to discuss the refinement of scientific measurement. Since many linear standards were end bars that had become worn or damaged, these geodesists considered it imperative that new standards be manufactured to reflect new scientific improvements. Once this was accomplished, all base measurements could be referred to the same linear standard, thus insuring that all European geodetic work was comparable. This would also enable scientists to determine a degree of a great circle of the earth with greater accuracy, since they would employ a number of different measurements. In order to secure such accuracy, the Convention recommended the construction of a new international prototype meter, differing in length as little as possible from the Meter of the Archives. An international commission would supervise the manufacture of this new standard and, in the interim, a bureau of weights and measures would be established to store it and to sponsor further improvements. This would make the metric system international in character; its preservation and development would be the concern of the world and not just France.

The action of this geodetic society encouraged other groups, such as the St. Petersburg Academy of Sciences, to petition the French Academy in 1869 to establish some common denominator by which an international metric system could be set up. The Parisian scientists did not receive these petitions enthusiastically, since they believed that other nations were attempting to diminish the French contribution to the metric system. (Nationalism triumphed momentarily over world concerns and common sense.) But the French government hesitated only for a short time, and upon the urging of the Academy, decided to take up the matter. After an examination of the question by a committee consisting of Academicians and Bureau of Longitude personnel, Alfred Leroux, the Minister of Agriculture and Commerce, issued a report supporting the plan and requesting that an international conference be formed to handle the specifics.

Emperor Napoleon III approved the report, and the government, through diplomatic channels, invited foreign governments to send delegates to a conference at Paris to discuss the construction of a new prototype meter and of secondary standards for the participating nations.

In August, 1870, delegates from 24 nations met at Paris. Since September, 1869, the French members had been working on the activities for the conference, and they decided that a number of identical standards should be constructed. One of these would be selected as the international primary standard and deposited in some convenient place accessible to all and under common supervision. The rest would become secondary standards and distributed to the delegates. The outbreak of the Franco-Prussian War, however, cut short the Conference, and the proposals reached were tabled for the time being.

Summoned anew by the French government in September, 1872, 30 nations sent 51 delegates. The first announcement made by the French committee stated that a detailed examination of the Meter of the Archives had been made and compared with those of the Conservatory and the Observatory. They found the state standard to be very well preserved, thus making it ideal for establishing the new international standard.[3] The Kilogram of the Archives also passed the inspection handily.

Eleven committees emerged to study and report on such diverse considerations as the condition of the ends of the Meter of the Archives, the material to be used in the construction of the new meter, together with its form and method of support, the temperature of the meter and kilogram when tested for official purposes, the creation of an international bureau of weights and measures, the weight of a cubic decimeter of water, and the methods to be used in preserving the standards.

After intense committee hearings, major decisions emerged quickly. One group decided to reproduce the Meter of the Archives by a line bar (mètre à traits), since they found that the ends of the platinum meter were accurate enough to warrant its use. They also advised that the other copies be line bars, but that a number of end bars (mètres à bouts) be constructed for nations requesting metric standards in the future. These bars must represent a true meter at 0° centigrade, the material being an alloy of 90 percent platinum and 10 percent iridium with a tolerance of error of no more than two percent. The bars, each 102 centimeters in length, must also be constructed from a single ingot produced at one casting and carefully annealed. Finally, the members provided detailed instructions for determining the expansion, the marking, and the calculation of the equations of the different bars.

The group assigned the kilogram declared that the standard of the Archives was exact enough for the purposes of industry and commerce, and even for most of the ordinary requirements of science. To change its mass at this late date would negate some important scientific work conducted since 1799 and produce needless expense. They decided to establish an international kilogram based on the weight of the Archives' standard when weighed in a vacuum, employing the same platinum-iridium alloy used for the meter standard. Also, it was to be identical in form to the previous standard. Eventually, the committee delineated the method to be followed in weighing and determining the volume of these kilograms.

The French scientists received the task of constructing these new standards, tracing their defining lines, and comparing them with the standards of the Archives, as long as the operations were under the general supervision of a permanent international committee of twelve members.

The commission also advocated the founding of an international bureau of weights and measures located at Paris that would be supported by contributions from nations signing the treaty. They proposed that the Bureau be under the supervision of a permanent committee of the International Metric Commission, and be used for the comparison and verification of new metric standards, for the custody of the new prototype standards, and for any other comparisons of weights and measures that might arise in the future. In accordance with the suggestions of the Commission, the French government communicated diplomatically once again with various governments on this plan, and when they submitted their responses, the participating nations signed a treaty in Paris on May 20, 1875 in which these recommendations were actualized. Soon thereafter, the United States, Germany, Austria-Hungary, Belgium, Brazil, Argentina, Denmark, Spain, France, Italy, Peru, Portugal, Russia, Sweden and Norway, Switzerland, Turkey, and Venezuela signed the treaty into law.

Of the countries present, Great Britain and Holland declined to participate in the treaty or to contribute to the expenses of an international establishment of the metric system. The Dutch offered no official explanation. The British government stated that they could not recommend to Parliament any expenditure in connection with the metric system since it had never been legalized in Great Britain as the sole basis of weights and measures, nor could it support financially a foreign institution. For reasons unknown, a change of feeling took place, because in September, 1884, Great Britain joined the Convention.

The treaty, providing for the establishment and maintenance of a permanent international bureau, specified that the signatory nations support on an equal basis a special building, supplied with appropriate instruments and apparatus, and staffed with scientists, a director, assistants, and workmen.[4] The first duty of the Bureau was to verify the new international metric standards. In future years it would retain custody of the international metric prototypes, conduct all official comparisons with those of the national standards, make comparisons between metric and non-metric standards, standardize geodetic instruments, and undertake metrological operations to supply the greatest benefit to supporting nations.

Standards were made, copies distributed, and the international bureau was established. The century-old dream of the French for a metric world was on its way to becoming a reality.

British Reaction and Retrenchment

The metric system spread rapidly following these pioneering scientific and diplomatic advances. Its growth and dissemination during the nineteenth and twentieth centuries followed definite patterns throughout the world. In the period before 1850, twelve nations introduced or adopted the system—the majority of them either shared a common frontier with France or had long-standing political and diplomatic relations with her.[5] These years were also marked by the start of the South and Central American conversion movements.[6] Between 1851 and 1900 metrication advanced at its most rapid pace—forty-six nations opted for it. This era witnessed the completion of the western European and South and Central American conversions, and the beginning of those in eastern Europe, Africa, and the Far East.[7] Twenty-two nations joined the metric world during the first 50 years of the twentieth century, most of them located in eastern Europe and the Far East.[8] Since 1951, eleven additional countries have gone metric, with the Near East having the largest number.[9] Currently, fifteen foreign governments are in the final stages of adopting metrics, and about a dozen others are considering its possible use.

During the nineteenth century when the metric system was spreading at a hectic pace throughout Europe and other parts of the world, its acceptance in the British Isles suffered from a fierce anti-metric movement and from governmental recalcitrance.[10] Aside from refusing to cooperate with the earlier Talleyrand proposal of French, English, and American co-sponsorship of metrics, Parliament failed to give proper consideration to the Miller and Keith plans during the period when the imperial system was in its formative stages. Regardless, the desirability of a decimal system for weights, measures, and coins began to be felt even before the passage of the act of 1824, because Sir John Wrottesley brought such a scheme to the notice of Parliament in 1814. The steadily increasing number of metric supporters forced the government to appoint a commission in 1819 to consider the question, and among its members were the well-known scientists Young, Wollaston, and Kater. Though this group eventually spoke against adoption of the metric system, or any decimalization of imperial weights and measures, the cause continued to grow, and various spokesmen for British government, science, engineering, commerce, business, education, medicine, and law debated vociferously its merits and deficiencies.

Even before the negative report of the 1819 commission appeared, Parliament passed a resolution in 1816 requesting the Royal Society to undertake the task of comparing the imperial standard yard with the meter. From Paris the Royal Society received two secondary platinum standards that had been compared with the primary standard. One was an end bar that was exactly equal to the primary standard at the temperature of melting ice; the other was a line bar that at the same temperature was short by 0.01759 millimeter. Kater compared them with the Shuckburgh scale and the Parliamentary primary standard. He determined the true length of the meter to be 39.37079 English inches, a value eventually legalized by Parliament in the act of 1864 that permitted use of metric weights and measures.

Prior to the passage of the 1864 statute, the cause of metrication in the British Isles received a considerable boost from the celebrated London Exposition of 1851. Despite the fact that twelve nations already had adopted metric weights and measures, and that mathematicians, chemists, physicists, astronomers, engineers, economists, educators, and statisticians had long advocated its adoption, the system failed to gain acceptance among British commercial, financial, and agricultural elements. (These were the critical sectors, of course, since they controlled the bulk of the national economy.) The Exposition brought together thousands of merchants, industrialists, and businessmen from all over the world, and they witnessed the advantages accruing to national and international trade from a common, universal system of weights and measures that provided uniform sizes, standardized dimensions, and interchangeability of parts and equipment. At its conclusion, the Society of Arts, in a communication addressed to the Lords of the Treasury, launched the earliest intensive metric campaign by asking whether it would be possible to adopt a universal system of weights, measures, and coins to facilitate international industrial and commercial growth and exchange. This was the first official declaration by a government institution of support for a decimalized metrology. But the Society took no further immediate action, due to the fact that confusion arose in both the public and private sectors as to whether they advocated the metric system or simply the decimalization of imperial units. The Society never made this clear. Whether this was an attempt to ease an eventual metric changeover by engaging in a bit of obfuscation at this early date is impossible to determine. Perhaps the latter tactic was considered necessary given the lamentable history of government reaction to metric or decimal proposals.

Further metric support came following the Second International Statistical Congress, held in Paris in 1855. James Yates (1789-1871), a participant in the Congress and a member of the Royal Society, proposed the formation of an international association to advance the adoption of a world-wide decimal system of weights, measures, and money. Once established under the name of International Association for Obtaining a Uniform Decimal System of Measures, Weights, and Coins, the British branch made a rigorous examination of the different monetary and metrological systems employed around the globe, and opted ultimately for the metric system because it had superior physical and scientific standards, it was immediately available for international trade purposes, and it had a non-nationalist nomenclature. In 1860 the Association sponsored a series of lectures on the subject of international metrology as a further means of advertising the advantages of metrics.

Another event occurred in 1860 that had a pronounced impact on this metric campaign—the signing of a commercial pact between Great Britain and France known as the Cobden Treaty. It marked the culmination of the free-trade movement in the British Isles, launched France in the same direction, and led to a series of "most-favored-nation" treaties among certain European nations. Since this and later diplomatic compacts produced a marked expansion in commercial relations throughout western Europe, and since several participating nations had already gone metric, British metric advocates obtained a powerful economic argument to use in trying to convince the government of the urgency of their cause. Soon thereafter in 1861, the Annual Conference of Deputies of the Associated Chambers of Commerce of the United Kingdom passed a resolution advocating the metric system, as did the National Association for the Promotion of Social Science.

These events led a Select Committee of the House of Commons in 1862 to consider a metric proposal and, after extensive hearings, to report that it would prove far easier to adopt the metric system than to go through the enormously arduous and expensive process of constructing an English decimal scheme in the hope of gaining eventual international acceptance as proposed by the English Decimal Association in 1858. Leone Levi (1821-88), a distinguished economist, testified before the committee that Germany opted for the system in 1861, and that other nations planned to make a similar changeover. He dwelt on the advantages of metrics to industry and education, but made the terrible mistake of urging the use of English names for all metric units.[11] This awkward alteration, that would have hindered severely British international commercial dealings, disturbed other witnesses greatly. In addition, they argued that any substitution of customary unit names for the international nomenclature would simply add to popular confusion and defiance. After several warnings from members cautioning the government about the expenses involved, Rowland Hill (1795-1879), Secretary of the Post Office of Great Britain and inventor of the adhesive postage stamp, while not advocating specifically the French system, pleaded for uniformity among different nations to facilitate a better postal arrangement, since Britain was losing large amounts of money on its postage to metric nations, even though the records of the proceedings never make clear how this happened or could happen. Yates, the promoter of the International Association, gave the most forceful pro-metric argument, however, by insisting that England was a leader in world industry and technology, and that other countries had postponed metric conversion to await the British decision. Dozens of other witnesses voiced similar concerns. The report of the Select Committee, dated July

15, 1862, followed these pro-metric recommendations closely.

The committee opted for the metric system as the only legal basis of weights and measures in the British Isles, for a newly created department of weights and measures, for metric employment in customs, for metric competence among civil service employees, for the use of the gram in the Post Office, for metric education in schools, for metric statistical usage, and for metric expression in all public documents.

In its final advice to Parliament, the British branch of the International Association pleaded for obligatory adoption of metrics, and against permissive use, for several historically verified reasons. First, the unfamiliar Graeco-Roman nomenclature caused enough public confusion in other countries. Why create a government-sponsored means of avoiding them in the United Kingdom? Permissive systems traditionally encouraged non-compliance. Further, when metrics were optional, those segments of the population who incurred financial expenses avoided changing over. (The groups implied here were the industrial, financial, and agricultural sectors.) Teaching both systems in the schools impeded progress in learning either the metric or the traditional system. Finally, permissiveness usually led to postponement of metric adoption for many decades. Britain faced the possibility of eventually lagging behind the rest of the world in technology and commerce if it hesitated at this strategic point in time. (Perhaps a stronger argument would have been the eventual loss of sales revenues.) Many scholarly and popular articles followed advocating the immediate obligatory adoption of the metric system.

Several other popularly sponsored bills, by such agencies as the British Association for the Advancement of Science, were brought before Parliament shortly thereafter, but a complete metric adoption failed to win enough of a government commitment. A bill of May 12, 1863, had only one reading and died. On July 1, a bill for metric adoption actually passed Commons by a vote of 110 to 75, but this session of Parliament was too far advanced to carry it to a vote in the House of Lords. The strict schedule of Parliament prevented a possible British metric adoption. Clock time, eliminated by the French in the formative metric years for incorporation into the system, proved the undoing of the English acceptance of metric weights and measures. Naturally, there may have been more than time to be considered here; perhaps the unfortunate scheduling signalled its ultimate doom. Never before was Britain closer to a metric metrology than in 1863. The Act of 1864, that simply permitted the system's use in the British Isles, was the government's half-hearted response to these significant events.

Parliament passed the Metric Weights and Measures Act of 1864 primarily to promote national and international trade, and to advance the multi-faceted interests of science.[12] Now, all contracts employing metric units were legal, and considered to be on an equal standing with those using imperial units. In the schedule attached to this act, the following imperial equivalents became legal throughout Great Britain and Ireland. The spellings, the metric values, and the unusual imperial equivalents provided are faithful to the letter of the act.

LINEAR

myriameter = 10,000 m = 10,936 yd, 11.9 in

kilometer = 1000 m = 1093 yd, 1 ft, 10.79 in

hectometer = 100 m = 109 yd, 1 ft, 1.079 in

dekameter = 10 m = 10 yd, 2 ft, 9.7079 in

meter = 1 m = 1 yd, 3.3708 in

decimeter = 0.1 m = 3.9371 in

centimeter = 0.01 m = 0.3937 in

millimeter = 0.001 m = 0.0394 in

AREA

hectare = 10,000 sq m = 11,960.3326 sq yd

dekare = 1000 sq m = 1,196.0333 sq yd

are = 100 sq m = 119.6033 sq yd

centiare = 1 sq m = 1.1960 sq yd

CAPACITY

kiloliter = 1 cu m = 3 qtr, 3 bu, 2 pk, 0.77 pt

hectoliter = 0.1 cu m = 2 bu, 3 pk, 0.077 pt

dekaliter = 0.01 cu m = 1 pk, 1.6077 pt

liter = 0.001 cu m = 1.76077 pt

deciliter = 0.0001 cu m = 0.176077 pt

centiliter = 0.00001 cu m = 0.0176077 pt

WEIGHT

millier = 1,000,000 g = 19 cwt, 5 st, 6 lb, 9 oz, 15.04 dr

quintal = 100,000 g = 1 cwt, 7 st, 10 lb, 7 oz, 6.304 dr

myriagram = 10,000 g = 1 st, 8 lb, 11.8304 dr

kilogram = 1000 g = 2 lb, 3 oz, 4.3830 dr

hectogram = 3 oz, 8.4383 dr

dekagram = 10 g = 5.6438 dr

gram = 1 g = 0.56438 dr

decigram = 0.1 g = 0.056438 dr

centigram = 0.01 g = 0.0056439 dr

milligram = 0.001 g = 0.00056438 dr

This particular intensive campaign was over, but the act of 1864 obviously did not go far enough to satisfy metric advocates since it merely accepted the French system as an equal partner in legal contracts. Another strong objection was that metric standards were not really legal because they had no existence in the law. Hence, use of the metric system was legal only on the books, but still not in practice. Pro-metric forces wanted its complete adoption and the elimination of the imperial system from all aspects of British life. They desired an end to more than a millennium of British metrological history.

Metric advocates readied their forces. Even before the act of 1864 was printed and became law, they launched another campaign to instruct all segments of the public on the advantages of metrics; they demanded its free and open competition with imperial weights and measures. After a member of the Institute of Mechanical Engineers delivered a paper in Birmingham late in 1864 espousing the theme of English weights and measures for Englishmen (a familiar battle-cry for anti-metric spokesmen, especially those who were most affected, either professionally or monetarily, by the new system), the International Association sent a delegation, including Yates, Levi, and Alexander Siemens (1826-1904), to argue the metric viewpoint. The fight became even more heated when the British Association for the Advancement of Science took over the role of metric propagandist from the International Association, after the latter became less active following substantial metric victories throughout Europe.[13] The Science Association commenced a series of programs endorsing metrics, that publicized how the recent act added to popular bewilderment and made metrological fraud more likely. After subsequent meetings over the next few years, the Association issued metric charts to help explain the system to the people, and posted metric materials in prominent places such as public squares, markets, business establishments, and meeting halls. (Once again, they were addressing the wrong audience; the French learned this lesson much earlier.)

In February, 1867, a conference was held in London of Chambers of Commerce that promoted the use of metrics. The deputies passed a resolution urging that metric weights and measures be introduced in commerce, industry, and the trades. To help facilitate this program, they mounted a campaign in the city of Paris, where an exhibition took place later in the year. After the exhibition, an International Committee of Weights, Measures, and Moneys passed a series of resolutions that so impressed the British government, that it issued a special report advocating metric adoption during 1868. Part of the campaign was the manufacture and distribution of mural or wall standards to help British citizens become familiar with the system. (Later metric advocates would learn that such methods usually antagonized the populace rather than drawing them closer to amiability.) Another step was to urge Parliament to draft a new bill providing for compulsory adoption.

The British government's final move was to appoint a group of Royal Commissioners to inquire into the condition of the Exchequer standards and to undertake an investigation of the whole subject of weights and measures. The group consisted of Chairman Airy, Edward Sabine (1788-1883), Graham, Miller, and H.W. Chisholm, Warden of the Standards and Superintendent of the Standards Department of the Board of Trade. On May 13, with the tacit consent of the government, the House of Commons received a new bill that advocated the repeal of the act of 1864 and the compulsory adoption of the metric system within a time deadline determined by the government. After a second reading, this bill of 1868 passed Commons by a vote of 217 to 65. But a delay in the receipt of the commissioners' report (how fortuitous these delays had become), the impending dissolution of the House, and the urgency of dealing with another non-metrological bill, led to its withdrawal on July 1. (One cannot blame Airy for this defeat.) This was a most unfortunate series of events. It represented another setback, but the campaign continued.

In April, 1869, a second report of the Standards Commission appeared that discussed the impact of metrics upon regional and international commerce. In regard to the first, the commissioners gathered data on metric adoptions in other countries, and concluded that the elimination of customary systems had no bearing whatever on internal commerce, either in the products traded or in their level of distribution. They emphasized that many countries of the world were going metric, especially industrial nations. Regretfully, they saw no objection to allowing permissive use of the metric system, since certain elements of society eventually would convert anyway out of necessity or community pressure. As to foreign trade, they observed that it was small in quantity compared with the level of internal exchange. Nevertheless, international commerce was of such growing magnitude that metrics should be permitted so that Britain´s weights and measures agreed with those of its foreign trade allies. (In later years foreign competition was the major consideration.)

Even though there was an unfortunate return here to the theme of metric permissiveness, the commissioners advocated that metric standards be deposited in the Standards Department, that they be legalized, and that verified copies be provided by local authorities for inspectors. The commissioners justified the reintroduction of the permissive clause on the grounds that compulsion hindered, rather than aided, adoption. (The French government found the opposite to be true in the period before 1837.) People, they felt, must want to change; coercion only led to reluctance and entrenchment. They believed that the metric system should be introduced into Great Britain very cautiously and over an extended period of time. (History proved such a course of action fatal in the past; future events would prove it so once again.)

The report submitted by the commission immediately antagonized the pro-metric forces since they thought that their entire program was threatened by revisionism. To them, the commissioners endorsed British metrological diversity, supported inferior imperial standards, and, most damaging of all, popularized the belief that customary units met the needs of British citizens as well as metrics.

Two positive results accrued from the second report, however. It highlighted class differences in regard to metric sponsorship by showing that pro-metric advocates came almost exclusively from the middle and upper classes; excluding, of course, the familiar industrial, financial, and agricultural elements. The lower, or "working" class, most affected by a changeover, opposed it or was indifferent to it; notwithstanding the many posters and murals aimed in their direction. Also, it demonstrated that metric conversion always worked best in countries where governments were repressive or authoritarian. Such political totalitarianism forced the masses to accept metrication. (The experience of the revolutionary regimes of France is a testimony to this conclusion.) In Great Britain the lower class was accustomed to expressing its views. Hence, adoption was much more difficult here than elsewhere.

The 1869 report crippled this particular metric endeavor, and put a damper on agitation for metric adoption. Not only did this current movement subside rapidly, but when it was revived later on it was not as radical in format or application. Other metric plans followed during the 1870s, but none succeeded, including those of the Metric Committee of the Science Association, the International Decimal Association, and the Joint

Committee of the Central Chamber of Agriculture. It was not merely overt opposition, but general dissolutionment that killed these intensive efforts.

The opposition, however, was extremely formidable, since it was during this era that the anti-metric forces of the British Isles were most active. They convinced large portions of the population that the imperial system was convenient and adaptable to British society, that it had special units for special purposes, and that it resulted from more than a millennium of evolutionary growth (a special variety of Darwinian "natural selection"). They argued that Great Britain and the United States used it, and that both nations stood in the forefront of world industrial production. Further, the English system dominated international trade, manufacturing, and navigation. Finally, the imperial binary system of division was immensely convenient for almost any trade, profession, or purpose. In due time, they would create numerous other justifications for retaining the old system.

On the other hand, these anti-metric propagandists emphasized that the metric system produced a greater likelihood for error since the decimal point could be put in the wrong place. Metrics also involved more figures in its computations, and this again increased the probability of error. They believed that the meter was too long for practical use, and lamented the fact that it was not derived from some natural, human standard (more Darwinianism). Decimal division was too difficult to learn; it was unsuitable for practical purposes; and it was not applicable in circle and time measurement. They insisted that the base of the system was suspect and not correctly determined. It was too difficult to think in metrics, its nomenclature was derived from a foreign language, and its adoption would invalidate all land titles, engineering drawings, technical literature, deeds, contracts, and similar documents. The metric system was advantageous only to teachers and scientists; not to workers and dealers. Even those countries that adopted it found popular resistance too great an obstacle to overcome. Metrics created confusion in elementary education and increased the time needed to make mathematical calculations. Opponents foresaw an inordinately long conversion period since the habits of an entire culture were impossible to legislate away in a few years. Finally, metric conversion would produce enormous expense through the disorganization of industry, the discarding of machinery, tools, and instruments, and the restandardization of products. The system was already permissible; whoever wanted it could use it. They saw no benefits deriving from a policy of change simply for the sake of change.[14]

These anti-metric arguments helped to crush the latest metric endeavors, but metric sponsors, once again, reassembled their forces. Before they could muster a definitive program, Parliament passed the Weights and Measures Act of 1878. Among its many provisions discussed earlier was one that gave the Board of Trade authority to verify metric weights and measures for science and industry or for other lawful purposes. Soon after its publication the new metric onslaught began. The Act of 1878 gave the metric system more of a legal footing in the British Isles than ever before. This spurred the cause forward even though only science and industry received direct metric sponsorship. Further aid came in 1884 when Great Britain signed the Metric Convention and agreed to help support the International Bureau. This was probably the psychological catalyst that the pro-metric forces ultimately needed, because the pace of their activities accelerated. They were determined to show British citizens the superiority of metrics by stressing its structure, denominational prefixes, meter-based standard, decimal divisions, interunit relationships, absence of product variations, universality, scientific underpinning, and educational, commercial, financial, manufacturing, and business advantages.

In the late 1880s, the metric cause received other inducements. Since more and more nations were going metric, this affected British trade and influenced even the most conservative British spokesmen. The formation of the International Bureau of Weights and Measures, the International Committee, and the various International Conferences helped to keep metrics before the public. The International Postal Union adopted the system, as did the International Congress of Engineers. (Oddly enough, many British engineers and engineering societies were among the most rabid anti-metric forces.) Finally, the United States gave signs of favoring metrics over its customary, pre-imperial, English system.

Because of these events, in the summer of 1887 a delegation of government, banking, and mercantile members urged the Chancellor of the Exchequer to adopt a decimal system of weights, measures, and money, and to consider metric employment. Although it is not known what precisely happened to this petition, several noteworthy advances were made shortly thereafter. In 1891 metrics began to be taught in British schools as part of the mathematics curriculum. In 1892, the annual Trade Union Congress of 495 delegates, representing more than a million members, passed a resolution favoring metrics. They repeated this action in 1893. In addition, the Decimal Association conducted many polls between 1890 and 1895 that showed a steady rise in metric support.

Such a favorable response arose that in 1895 the government appointed a Select Committee to conduct hearings. Under the chairmanship of Henry E. Roscoe (1833-1915), a chemist, this committee, drawn from both Commons and Lords, listened to testimony from experts in industry, commerce, government, banking, education, and other professions. H.J. Chaney, Superintendent of Weights and Measures, stressed that metrics had made little inroad into Britain since not one of the three million weights stamped by his department was metric. A witness from the Institute of Civil Engineers claimed that the imperial system was inimical to further commercial growth. After reciting the names and opinions of numerous pro-metric sponsors, he emphasized that British commerce needed a metrological change. Siemens, the past president of the Institution of Electrical Engineers and a member of the Council of the Society of Arts, stated that his company converted to metrics around 1870 and that his employees considered the system a definite improvement. Sir Richard Strachey (1817-1908), an Anglo-Indian administrator, told of Indian gains from metric use. Alfred Spencer, head of the Public Control Department, called for as short a changeover period as possible since he considered the cost to retail merchants as minimal. P.J. Street, Chief Inspector of Weights and Measures for the City of London, spoke for wholesale dealers, bullion and diamond merchants, jewellers, and silversmiths who were pro-metric, and countered the arguments of the anti-metric sectors whom he considered to be ignorant of metrics. Representatives of clothing manufacturers, educational institutions, chemical firms, grain and seed distributorships, municipal governments, timber companies, pharmaceutical houses, engineering and technological concerns, export trade, and railroad and bridge factories urged metric adoption. Finally William Thomson, Lord Kelvin (1824-1907), the famous mathematician and physicist, spoke of the value of metrics to science, engineering, and land surveying.

On July 1, 1895, the Select Committee issued a report that endorsed metrics enthusiastically. Since more than 50 nations had adopted the metric system during the last half of the nineteenth century, they recommended that the metric system be legalized immediately for all purposes in the British Isles; that after a two year waiting period, the system be made compulsory by an act of Parliament; that metrics be taught in all elementary schools; and that decimals be introduced as part of the mathematics curriculum.

Parliament hesitated again because of the anti-metric movement, and acted only on that portion of the report that requested the legalization of metrics for all purposes. On May 27, 1897, Charles Thompson Ritchie (1838-1906), a member of Parliament and President of the Board of Trade, introduced a bill to this effect, and after a third reading on July 8, it passed. Thus, after many decades of struggle, metric weights and measures were now permissible for any use in Great Britain. Compulsory usage would have to wait many decades longer. On May 19, 1898, an Order in Council defined again the unit definitions of the system and provided their equivalents in imperial weights and measures.

Since no obligatory metric changeover occurred, metric supporters continued their efforts. They received encouragement in 1902 from a Conference of Prime Ministers of Self-Governing Colonies held in London. Opting in favor of metric weights and measures were the ministers of Canada, New Zealand, the Cape of Good Hope, Transvaal, the Orange River Colony, Southern Rhodesia, Gambia, Northern Nigeria, Gibraltar, British Guiana, Trinidad, the Leeward and Windward Islands, Sierra Leone, Southern Nigeria, and the Falkland Islands. Jamaica and British Honduras indicated the desire to go metric if the United States were favorably inclined. Fiji and British New Guinea left its decision up to Australia. Newfoundland, Malta, and Bermuda had to await the decision of Canada. The Straits Settlements deferred to India, while South Africa agreed to go whenever the Empire decided. Mauritius and Seychelles already used the system exclusively. The only opposition came from St. Helena, Cyprus, Lagos, Barbados, the Gold Coast, Queensland, and several small states. During this same period, many of the nation's journals and periodicals also published articles lauding these and other pro-metric efforts.

On the basis of these events, it seemed to the members of the Decimal Association that support of radical reform was increasing, and that the time was right to make an all out effort to adopt the metric system. The Association actively solicited the support of individual members of Parliament and secured written pledges from many of them. Another tactic was to get affirmative testimonials from hundreds of prominent British citizens. Accordingly, Lords Belhaven and Stenton, supported by Lords Kelvin, Rosebery, Spencer, and Tweedmouth, introduced a bill in 1904 advocating metric adoption. This plan provided for the establishment of the standard kilogram and meter from April 1, 1909, as the only British standards of weights and measures, though Parliament could delay this date if conditions warranted it. (Another tactical error made, perhaps, in the euphoria of the moment.) The sponsors requested that Parliamentary copies of metric standards be manufactured, and that future contracts be made solely in terms of the metric system. They advised Parliament to be careful to allow for staggered adoption deadlines by various sectors of the economy so as to lighten the economic burden of conversion. Finally, the bill prescribed the general methods by which the changeover should be carried out. Lords Belhaven and Stenton stressed to Parliament that metrics was very popular among the major cross sections of British society, and cited the endorsements of two Select Committees, the consular reports, the resolution of the Prime Ministers of the Empire, and the statements of some of Britain's most famous citizens. After a third reading, and after the bill was passed by the Lords, it was sent to Commons where, unfortunately, it failed to pass by a close vote of 150 to 126 even though it had endorsements from town, city, and county councils, over 50 urban chambers of commerce, 42 trade unions representing almost 5,000,000 members, 60 teachers´ associations, metrological inspectors in 80 districts, 30 retail trades´ associations, and numerous agricultural chambers and farmers´ associations. By 1906, the number of those voting in Parliament for a metric bill increased to 414, an increase of 318 since 1900 and a clear majority, but again in 1907 it was passed by Lords but killed in voting by Commons by only a few votes due to abstentions by some of the members who had previously signed pledges. It could certainly be argued that the political careers of the representatives in Commons were more dependent on the political clout of certain powerful lobbies than were those of the upper House. More than forty years would pass before the metric cause would arise strongly again.[15]

This latest metric drive expired as had so many previous ones. One reason for the failure was the influence of last minute anti-metric propaganda on the nation´s manufacturing interests, and the latter´s political impact on many parliamentary members. Wavering members of both Houses, but primarily Commons, feared angering any large or vocal segment of the population. The political repercussions were too great. Also, they weighed the views of some committee members who stressed objections to metrics, based on the total cost of conversion and the significant disruption of various industries and business concerns.

This was as close as Britain ever came to a metric changeover before the second half of the twentieth century. Although the 1907 bill lost by only 32 votes, metric sponsors felt severely defeated and betrayed. The

Decimal Association lost some of its membership and considerable revenue during the following decade, and neither was reinforced until the war years renewed British interest in metrication. Metrics had been before the public eye for a long time and it was impossible to maintain this intensity indefinitely. In the interim some discussion of adoption continued in Britain, and throughout the Empire periodical literature maintained a small but steady stream of articles devoted to metrics. For example, the Eighth Congress of Chambers of Commerce of the British Empire endorsed it in 1912, followed a year later by the Bradford Textile Society and the Far Eastern Section of the London Chamber of Commerce. By 1917 metrics gained the support of the British Horological Institute, the British Society of the Argentine Republic, the British Imperial Council of Commerce, the Conference of Irish Chambers of Commerce, the Court of Common Council of the City of London, the British Pharmacopoeia, the International Parliamentary Commercial Conference, the Conference of Scientific Societies, the British Institute of Bankers, and many other prestigious associations representing hundreds of thousands of members and associates. This popular and public concern enabled metric supporters to continue their efforts, however weak they may have been.

World War I brought new life to the metric cause owing largely to cooperation with metric system allies both before and during the conflict. Metric agitation increased, and this led to a revival of the Decimal Association. But little of substantial merit was realized save for strong and convincing arguments in the Decimal Educator by many distinguished citizens centering on the supposed massive endorsement of the system. They did get the backing by 1920 of the World Trade Club of San Francisco, an avowed metric supporter, and the American Metric Association of New York. They conducted polls, referenda, questionnaires, and other devices to arouse a governmental commitment. None was forthcoming. The Decimal Association continued, even though its journal ceased publication in 1936 due to insufficient funds. The metric well had run dry, at least for the time being.

There was only sporadic interest in metrication after World War I. More associations opted in its favor, together with the accustomed plethora of petitions, pamphlets, booklets, reports, endorsements, meetings, and advertisements. Global conditions put other pressing concerns on center stage. Great Britain had to await the end of World War II before another major metric drive could begin.

8
THE FINAL VICTORY

There was no British legislation mandating the use of metric weights and measures following the intensive campaigns of the late nineteenth and early twentieth centuries. The law concentrated solely on the imperial system, and aside from several acts, parliamentary regulations, and amendments during the first half of the twentieth century, the act of 1878 constituted the foundation of that law until 1963.[1] During this period, however, there were so many radical changes in various sectors of the economy that an updating of metrological law became imperative. For instance, there was enormous growth in the transportation and distribution of consumer goods and in marketing techniques. Technological and computerizational improvements caused a mechanical revolution. There were larger commercial, industrial, and mercantile conglomerates, and a corresponding increase in the volume and variety of goods produced for retail and wholesale trade. Thus, the government needed to provide a new weights and measures law and additional, modernized training methods for its weights and measures inspectors.

To prepare for this metrological overhaul, the government, during the late 1940s, empanelled a committee of Board of Trade members and requested that it make a comprehensive survey of British weights and measures and issue a series of guidelines for future action. Under the chairmanship of Sir Edward Hodgson, the Committee issued a report in May, 1951, that advocated the eventual abolishment of imperial weights and measures in favor of the metric system. The members believed unanimously that the metric system was superior to the imperial, and that sooner or later Britain would be forced to adopt it. Proceeding cautiously, the Hodgson report urged that discussions of metric conversion take place as soon as possible with leaders of industry and commerce to determine a reasonable timetable for transition. Calling upon the Commonwealth to strike an agreement with the United States for a simultaneous conversion, since these countries were the major trade partners of Great Britain, it recognized that the government must prepare the general public carefully, but expeditiously, for this significant alteration of British custom. It was also necessary to issue thorough and detailed plans during the transition period to assist those segments of the economy severely affected, even to the point of financial compensation whenever necessary. Their final recommendation was that British coinage be restructured on a decimal scale. (This was more than a century and a half after America's decimal monetary conversion, but the United States still lags behind Britain's metrological conversion.)

American Recalcitrance

The metric cause was alive again. Not since the campaigns antedating World War I did an official panel appointed by the government opt for metrics. This time metrication would be successful, but the hoped-for simultaneous conversion with the United States would never occur, even though American programs advocating metric adoption increased dramatically following World War I. For example, in 1919 the World Trade Club began operations in San Francisco, and for the next two years issued a barrage of pro-metric pamphlets while retaining a representative in Washington, D.C. to lobby for congressional support.[2] By 1920 they had collected 100,000 petitions endorsing metric adoption and submitted them to the government. In 1921 a subcommittee of the Senate Committee on Manufactures held extensive hearings on a metric adoption bill but, as usual, no legislation resulted. The strongest case of the era was made in 1922 by Aubrey Drury, a member of the World Metric Standardization Council, in World Metric Standardization: An Urgent Issue. Additional Congressional hearings followed in 1926, 1937, 1957, 1958, 1959, 1960, 1961, 1963, and 1965. Despite the fact that a formal law—the Metric Conversion Act—was finally enacted in 1975 proposing a ten year changeover, defeat of metrics was almost a foregone conclusion.

During the nineteenth and early twentieth centuries, American resistance to metrics followed many of the same patterns and themes found in the British Isles. By the second half of this century, however, American recalcitrance was based on the following arguments. First, since the industrial, agricultural, financial, and technological development of the United States had progressed after the Civil War at such a rapid and dramatic pace, the adoption of the metric system would curtail, or even destroy, further economic growth. Proponents of the American customary system contended that all available evidence, based upon the experience of other countries, indicated that the substitution of metric designations for existing sizes, and the actual replacement of American with new metric equipment, was impractical, and that if the United States contemplated a change in systems it must face the destruction of existing mechanical standards. In addition, they insisted that, following the change, there would be a long aftermath during which the mechanical industries of the country would suffer from tremendous confusion and the laborious undertaking of rebuilding new standards in another system.

Second, metrication would be too expensive a venture due to the aforementioned reasons. Because a compulsory change to the metric system would entail the discarding or alteration of a large part of the basic mechanical equipment of the manufacturing industries of the country, would compel the replacement of scales and measuring instruments in use among all classes of people, and would require a period of training in the use of the new system, the only result of a compulsory adoption would be to drag the country into an enormous expenditure and waste without providing any compensatory advantages. In fact, by the 1960s, even metric proponents were citing costs of twenty billion dollars, while those opposed used figures of one hundred billion dollars and higher.

In conjunction with this, the customary system had proved its overall superiority due to the fact that the United States had achieved a phenomenal growth rate in its Gross National Product since the Civil War, and that such growth was linked inseparably to the use of English weights and measures. This is the weakest argument any dissenting group or society ever offered in their opposition to metrics. It is tantamount to basing American economic dominance on the fact that English is the national language.

Further, customary metrology had become an integral part of American culture, and to adopt a foreign system would seriously disrupt the social life of the nation. Fringe elements even charged that metrics would weaken the moral fabric of American life. Not even the most fanatical British anti-metric group ever proposed such a ridiculous claim.

As far as its structure was concerned, a decimalized system was not viewed as superior, intrinsically or extrinsically, to the old system. In fact, the customary system was considered more adaptable to the intricacies of the American economic machine since it allowed for greater variances in product dimensions, designs, manufacturing techniques, advertising schemes, and the like. Its superiority was apparent also because its fundamental units, such as the inch, foot, pound, ton, quart, and gallon, developed from the eternal process of natural selection of the most appropriate units, and not as the result of a rigid, inflexible, and uncompromising plan. As in the British Isles, Darwinianism was used even by the most unsuspecting of individuals and groups.

Other reasons mentioned frequently were that Americans had gotten too accustomed to using fractions, and that decimals were not as adaptable or as easily understood in popular parlance and everyday activities. Since metrication would require a total national effort, other more pressing matters of national concern would be threatened, or even ignored outright. Some maintained that the change would involve enormous disturbances in the habits and customs of the public and private sectors to the overall detriment of the nation's well-being. Others argued that the investment of huge sums of money, both initially and over the long-run, would not warrant the slight increase in returns from foreign sales. Eventually the American manufacturing and production complex would lose its unique individual personality, and would make and distribute goods that would vary only in limited degrees from those of foreign competitors. (An analysis of historical import and export statistics renders such a claim to be totally unfounded.)

Finally, resistance to metrological change was so nationally and indelibly ingrained in the American way-of-life that the metric system would falter eventually through massive non-compliance with government directives. (Again, this runs counter to the actual experience of other countries.) English-system advocates believed that the ordeals endured by foreign nations showed conclusively that the difficulties arising from a drastic change in the habits of the people, from the necessity of revising the technical literature of the country, and from the confusion incident to the use of two systems side by side during the long period of transition necessary, would be insurmountable no matter what form of compulsory law was adopted. And since compulsory laws were required, the metric system had no advantage that would lead people to adopt it voluntarily. (The experience of more than a hundred nations since the early 1800s made this claim meaningless.)

These objections, however, did not influence negatively the British effort. With or without the United States, Britain was determined to resolve its own metrological conversion.

The Weights and Measures Act of 1963

Before all of the preliminary work entailed in a metric changeover plan could be completed, and with some facets of it hardly even begun, Parliament passed the Weights and Measures Act of 1963. This act was certainly not revolutionary, but several of its provisions helped to ease the eventual metric commitment. Having received royal assent on July 31 of that year, it simply amended, consolidated, and superseded many earlier pieces of legislation. It kept the substance of the older law but streamlined its form, especially in the areas of weights and measures standards and the procedures to be followed for examination and verification. The Board of Trade remained the directing ministry, enforcement still belonged to local authorities who appointed inspectors, and the old pattern of over 250 authorities in Great Britain was left unchanged.

The only significant realignment brought about by the act concerned units and physical standards. In the case of the latter, there were new definitions for linear and weight standards. Since the United Kingdom primary standard yard, as well as its four copies of 1845 and one from 1879, were of questionable workmanship, especially their defining lines and supports, and since they had deteriorated at a much faster rate than the international prototype meter, the 1963 statute defined the yard for the first time in terms of the meter (see Figure 17).[3]

This decision was based on two facts. First, the international prototype meter is a far more accurate standard. Second, it has been based

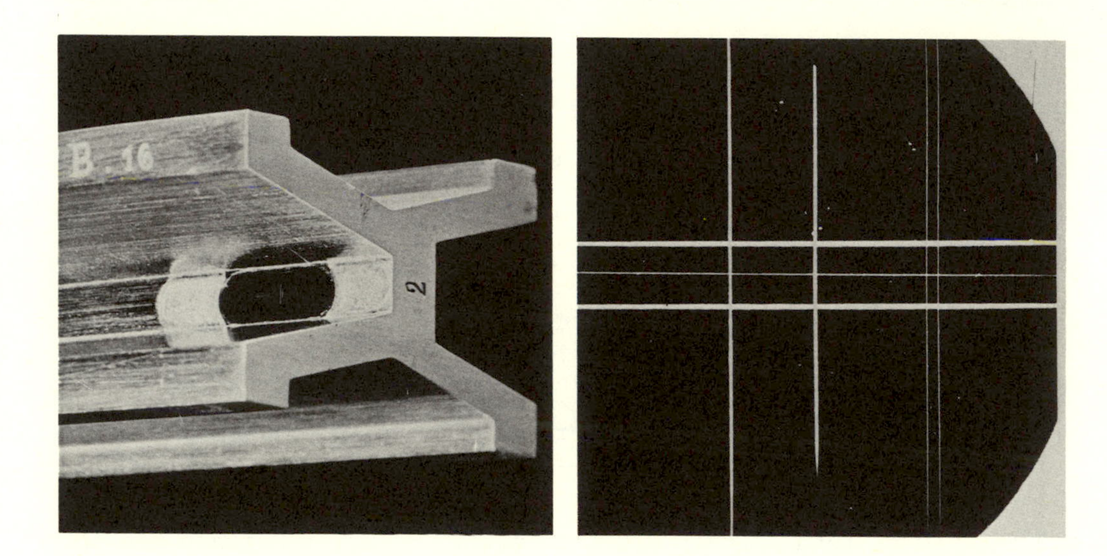

(a) (b)

Figure 17.

(a) End view of the British copy of the former International Meter showing the polished facet on the neutral plane of the bar.

(b) Microphotograph of the rulings on the polished facet; the finer graticule lines were used for visual settings by micrometer, while the two thicker vertical lines represented the expansion of the bar from $0^{o}C$ to $20^{o}C$.

(Photograph and copyright by John Moss. Furnished by the National Physical Laboratory, Teddington, Middlesex) since 1960 on the most scientifically accurate constant yet discovered, the wavelength of orange-red krypton 86.[4] The act also redefined the avoirdupois pound in terms of the international prototype kilogram, even though the imperial platinum cylinder of 1844 remained as the United Kingdom primary standard. The new definitions promulgated were: yard = 0.9144 meter exactly; meter = 1,650,763.73 wavelengths in vacuum of the radiation corresponding to the transition between the levels $2p_{10}$ and $5d_5$ of the krypton 86 atom; pound = 0.45359237 kilogram exactly;[5] and kilogram = the unit of mass represented by the mass of the international prototype kilogram. These were the only units by which measurements of length or weight could be expressed in the United Kingdom. The linear standards, being defined by a reference to a natural atomic standard, would remain unchanged until a new constant emerged through scientific research, but the weight standards, based upon a particular piece of metal, would be subject to very small changes in value over the years.

Based upon these definitions, Parliament authorized the use of the following units in the British Isles.

Table 8.1

Imperial and Metric Units Authorized in the Act of 1963

IMPERIAL LENGTH		METRIC LENGTH	
mile	= 1760 yards	kilometer	= 1000 meters
furlong	= 220 yards	decimeter	= 1/10 meter
chain	= 22 yards	centimeter	= 1/100 meter

Table 8.1 (continued)

foot	= 1/3 yard	millimeter	= 1/1000 meter
inch	= 1/36 yard		

IMPERIAL AREA

square mile	= 640 acres
acre	= 4840 square yards
rood	= 1210 square yards
square yard	= a superficial area equal to that of a square each side of which measures 1 yard
square foot	= 1/9 square yard
square inch	= 1/144 square foot

METRIC AREA

hectare	= 100 ares
dekare	= 10 ares
are	= 100 square meters
square meter	= a superficial area equal to that of a square each side of which measures 1 meter
square decimeter	= 1/100 square meter
square centimeter	= 1/100 square decimeter
square millimeter	= 1/100 square centimeter

IMPERIAL VOLUME

cubic yard	= a volume equal to that of a cube each edge of which measures 1 yard

Table 8.1 (continued)

cubic foot	= 1/27 cubic yard
cubic inch	= 1/1728 cubic foot
METRIC VOLUME	
cubic meter	= a volume equal to that of a cube each edge of which measures 1 meter
cubic decimeter	= 1/1000 cubic meter
cubic centimeter	= 1/1000 cubic decimeter
IMPERIAL CAPACITY	
gallon	= the space occupied by 10 pounds weight of distilled water of density 0.998859 gram per milliliter weighed in air of density 0.001217 gram per milliliter against weights of density 8.136 grams per milliliter
bushel	= 8 gallons
peck	= 2 gallons
quart	= 1/4 gallon
pint	= 1/2 quart
gill	= 1/4 pint
fluid ounce	= 1/20 pint
fluid dram	= 1/8 fluid ounce
minim	= 1/60 fluid dram
METRIC CAPACITY	
hectoliter	= 100 liters

Table 8.1 (continued)

liter	= shall have the meaning from time to time assigned by order of the Board of Trade[6]
deciliter	= 1/10 liter
centiliter	= 1/100 liter
milliliter	= 1/1000 liter
IMPERIAL WEIGHT	
ton	= 2240 pounds
hundredweight	= 112 pounds
cental	= 100 pounds
quarter	= 28 pounds
stone	= 14 pounds
ounce	= 1/16 pound
dram	= 1/16 ounce
grain	= 1/7000 pound
troy ounce	= 480 grains
troy pennyweight	= 24 grains
apothecaries ounce	= 480 grains
apothecaries dram	= 1/8 apothecaries ounce
apothecaries scruple	= 1/3 apothecaries dram
METRIC WEIGHT	
ton	= 1000 kilograms
quintal	= 100 kilograms

Table 8.1 (continued)

hectogram	= 1/10 kilogram
gram	= 1/1000 kilogram
carat	= 1/5 gram
milligram	= 1/1000 gram

The bushel, peck, and pennyweight ceased as legal units on January 31, 1969, and apothecaries weights and measures were affected similarly later on. The metric carat was reserved strictly for transactions in precious stones or pearls, and the troy ounce for transactions in, or articles made from, gold, silver, or other precious metals. Eliminated were three linear units formerly included in the Weights and Measures Act of 1878: the rod, pole, or perch of 5 1/2 yards (including the area measure of 30.25 square yards), the link of 6.6 feet, and the nail of 1/16 yard. Two capacity units were also discontinued: the quarter of 8 bushels and the chalder of 36 bushels. Areas of land had to be referred to in terms of acres and square yards, with the rood employed for appropriate areas between the two and the square mile for very large areas. The act introduced the cubic yard as a unit of volume for general trade use. A final change was the inclusion of the fluid ounce among the more familiar imperial units. Previously that had been primarily an apothecaries measure. Totally unaffected by the act were the "proof gallon" used in wholesale liquor sales as a quantity-quality measure; the "hoppus foot" important in timber sales for determining the quantity of usable wood in a tree trunk through employment of a specially calibrated tape; and "ship´s tonnage," or the estimate of a ship´s carrying-space, used for assessing harbor tolls.

The Metric Changeover

The metric debate continued unabated throughout the period encompassed by the Hodgson Report and the Act of 1963.[7] After this last imperial act, customary weights and measures still dominated, but their days were numbered. During the 1960s, surveys conducted by the British Standards Institution, the British Association for the Advancement of Science, and the Association of British Chambers of Commerce indicated that a majority of British industries favored an immediate change regardless of the course the United States eventually took, thus eliminating a major hurdle to conversion. (A wise choice indeed, considering the experience of past attempts at metrological cooperation.) In fact, in 1965, the president of the Federation of British Industries personally committed industry collectively to metrics, and requested that the government initiate a conversion program. Other inducements followed quickly. By 1968 over 75 percent of world trade was in metric terms. The emergence of the European Common Market and the Free Trade Association meant closer economic links, and stiffer economic competition, with an increasingly universal metric world. Because of these conditions, endorsement or acceptance of metrics, especially for exports, came from many diversified companies dealing in bulk materials, building supplies, construction, chemicals, paper, printing, iron, steel, non-ferrous metals, shipbuilding, electrical and non-electrical engineering, motorized vehicles, metal goods, precision instruments, jewellery, textiles, food, drink, and tobacco products. Government agencies, showing various levels of support, were the Board of Trade, Board of Customs and Excise, Ministry of Agriculture, Fisheries & Food, Her Majesty's Stationery Office, Ordnance Survey Office, Directorate of Overseas Surveys, Post Office, Ministry of Transport and Civil Aviation, Admiralty War Office, Ministry of Defense, Ministry of Supply, Central Statistical Office, Ministry of Health, Home Office, Ministry of Housing & Local Government, Inland Revenue, Ministry of Labor & National Service, Her Majesty's Land Registry, Ministry of Pensions & National Insurance, Ministry of Power, Scottish Home Department, Treasury, and Ministry of Works. Almost unanimous acceptance of the metric system came from local authorities, with the County Councils' Association, County and Non-County Boroughs, and Urban and Rural District Councils leading the way. Finally, other sectors of British society gave their approval, among the most important being the Gas and Electricity Councils and Boards, British Transport Commission, British Overseas Airways Corporation, British European Airways, National Coal Board, Atomic Energy Authority, United Kingdom Trade Commissioners, Royal Institute of British Architects, and many scientific, medical, legal, and statistical societies.

Such a massive outpouring of professional, governmental, and public support during the 1960s, regardless of the degree of commitment inherent in some of them, finally moved Parliament out of its historical doldrums and into affirmative metric action. Unlike past metric drives, this one was not led solely by decimal associations, scientists, educators, physicians, and pharmacists. There was an almost universal appeal to this campaign, that intensified following the Act of 1963, and led to the initial changeover steps taken by Parliament in 1965. In the period that followed, this same support sustained the metric transition during the early difficult and somewhat turbulent years.

On May 24, 1965, the President of the Board of Trade accepted a request from industry and made a statement to Commons urging the government to implement a carefully planned changeover to metrics. Parliament accepted the challenge—metric weights and measures were to be obligatory throughout Great Britain. A century of metric agitation came to fruition. Gone was the permissive metric status of past eras. Gone also was the customary system that received its initial standardization under King Edgar, more than a millennium earlier. One metrological era died; another was given birth.[8]

Parliament realized that the road ahead was bound to be rocky—a simple governmental directive would not be enough to change habits formed over such an enormously long period. Hence, both Lords and Commons recognized that the process of change must be gradual, and they proposed a time span of ten years for the industrial transition to the new system. More than two years went by, however, before the program could be launched. On May 14, 1968, the Standing Joint Committee on Metrication, with the support of the Council of the Confederation of British Industry, the British Standards Institution, the Royal Society, and the Council of Engineering

Institutions, implemented the 1965 statement. They recommended first that the government make the end of 1975 the termination date for the adoption of the metric system by the entire country, it being understood that those concerned would establish metric conversion programs to coincide with this date. If conditions warranted it, a later date would be permissible. Second, it was mandatory that the government establish planning bodies to identify the major problems and to prepare sector conversion programs. Third, a Metrication Board sponsored by a central department of government was necessary to oversee, stimulate, and coordinate the program initiated by the sector planning bodies. Acting as a central coordinating agency, the Board would prepare the general public for the changeover. Finally, it was incumbent upon Parliament to pass legislation, no later than January, 1971, explaining the specifics of the metric system and its application to all sectors of British society. The Board was a purely advisory body, and its members reflected the interests of industry, retail and wholesale trade, education, and the general public.

Soon thereafter various segments of Great Britain reported major gains in metrication, notably industry, construction, vocational training and retraining, pharmaceuticals, postal services, airlines, freight, shipbuilding, agriculture, transportation, communications, automotive industries, engineering, sea navigation, consumer trade, highway departments, and small business. Posters, exhibitions, advertising campaigns, local meetings, and study groups formed to help British citizens think in metric, rather than going through the laborious and tedious process of converting inches and pounds through arithmetical calculations. From January, 1969, all contracts for the Ministry of Transport, Housing Ministry, and Public Buildings and Works had to be in metric weights and measures. The Department of Education and Science required all building specifications submitted to be metric to qualify for grants. The Ministry of Defense preferred metric equipment after January, 1970. Industry realized a 50 percent increase in foreign contracts by March, 1970.

The final step occurred in 1971 when Britain decided to join the European Common Market. Since all the other nations in this consortium were totally metric, Great Britain had to be metric to compete economically on equal terms. After the publication by the British government in 1972 of the White Paper on Metrication, the pace of metric adoption accelerated, especially among smaller industries. There was now no doubt concerning the government's total commitment to a metric Great Britain.

There is still some lingering resistance to metrication in Great Britain, as there was in every other country since the early nineteenth century, but the campaigns of the last hundred years ended in success. British metrology is no longer dominated by insular concerns; it is part of a world community.

APPENDIX 1
PRINCIPAL DIFFERENCES IN IRISH AND WELSH WEIGHTS AND MEASURES COMPARED WITH ENGLISH STANDARDS

Unit	Description	English Standard	Deviation
IRISH LINEAR MEASURES			
Chain	land surveying: 1008.0 inches (25.60 m) or 100 links of 10.08 inches each	Gunter´s or Surveyor´s: 792 inches (20.116 m) or 100 links of 7.92 inches each	1.2727
Foot	English standard	12 inches (0.305 m)	none
Furlong	340 feet (2.560 hm) or 40 perches of 21 feet each	660 feet (2.012 hm) or 40 perches of 16 1/2 feet each	1.2727
Inch	English standard	1/12 foot (2.54 cm)	none
Link	land surveying: 10.08 inches (0.2560 m) equal to 1/100 chain	Gunter´s or Surveyor´s: 7.92 inches (0.2012 m) equal to 1/100 chain	1.2727
Mile	road distances: 6720 feet (2.048 km) or 2240 yards	5280 feet (1.609 km) or 1760 yards	1.2727
Perch	21 feet (6.401 m) or 7 yards	16 1/2 feet (5.029 m) or 5 1/2 yards	1.2727
Yard	English standard	36 inches (0.914 m)	none

Appendix 1

Unit	Description	English Standard	Deviation
WELSH LINEAR MEASURES			
Cyvelin	cloth in North Wales: 9 feet (2.743 m)	none	none
Foot	English standard	12 inches (0.305 m)	none
Inch	English standard	1/12 foot (2.54 cm)	none
Leap	land surveying: 6.75 feet (2.059 m)	none	none
Llathen	cloth: 9 feet (2.743 m)	none	none
Pared	cloth in Montgomery-shire: 9 feet (2.743 m)	none	none
Perch	English standard	16 1/2 feet (5.029 m) or 5 1/2 yards	none
Ridge	land surveying: 3 leaps (6.176 m) or 20.25 feet	none	none
Yard	English standard	36 inches (0.914 m)	none
IRISH AREA MEASURES			
Acre	plantation acre: 7840 sq yards (0.655 ha) or 160 sq perches of 7 yards each	4840 sq yards (0.405 ha) or 160 sq perches of 16 1/2 feet each	1.6198

Appendix 1

Unit	Description	English Standard	Deviation
Rood	1960 sq yards (0.164 ha) or 40 sq perches of 7 yards each	1210 sq yards (1.101 ha) or 40 sq perches of 5 1/2 yards each	1.6198
WELSH AREA MEASURES			
Acre	erw or standard acre in North Wales: 4320 sq yards (0.361 ha); originally the extent of land tilled in a day;	4840 sq yards (0.405 ha) or 160 sq perches of 16 1/2 feet each	0.8926
	stang or customary acre in North Wales: 3240 sq yards (0.271 ha)	same	0.6694
Bat	11 sq feet (1.022 sq m) in South Wales; originally the square of a large staff or club	none	none
Cantrev	25,600 erws (ca. 9241.60 ha) or 2 cymwds or 100 trevs	none	none
Cymwd	12,800 erws (ca. 1620.80 ha) in Anglesey or 50	none	none

Unit	Description	English Standard	Deviation
	trevs equal to 1/2 cantrev; originally any co-mote		
Cyvar	3240 sq yards (2709.063 sq m) in Anglesey and Carnarvon;	none	none
	2430 sq yards (2031.723 sq m) in Merionethshire;	none	none
	2821 sq yards (2358.725 sq m) or 192 llath of 11 1/2 sq feet each in South Wales	none	none
Erw	North Wales: See acre	none	none
Gavael	64 erws (ca. 23.10 ha) or 4 rhandirs; originally a customary, standard land-holding	none	none
Gwaith	peat in North Wales: 150 sq feet (13.935 sq m)	none	none
Llath	11 1/2 to 24 feet in South Wales (1.068 to	none	none

Appendix 1

Unit	Description	English Standard	Deviation
	2.230 sq m); originally the square of a large rod or staff		
Maenol	1024 erws (ca. 369.66 ha) or 4 trevs; originally signified an extent of rock-strewn land	none	none
Paladr	20.25 sq yards (16.929 sq m) in Anglesey	none	none
Rhandir	16 erws (ca. 15.78 ha) or 4 tyddyns; originally an extent of share-land	none	none
Stang	North Wales: See acre	none	none
Trev	256 erws (ca. 92.42 ha) or 4 gavaels; originally signified a vill	none	none
Tyddyn	4 erws (ca. 1.44 ha); originally the area of ground encompassing a homestead or tenement; similar to demesne-land	none	none

Appendix 1

Unit	Description	English Standard	Deviation
IRISH CAPACITY MEASURES			
Barrel	ale: 8704.0 cu inches (1.427 hl) or 40 gallons equal to 2 ale kilderkins or 4 ale firkins;	before 1688: 32 gallons (ca. 1.48 hl);	0.9642
		1688 to 1803: 34 gallons (ca. 1.57 hl);	0.9089
		after 1803: 36 gallons (ca. 1.66 hl)	0.8596
	barley and rape: 16 stone (101.604 kg);	none	none
	beans, peas, and wheat: 20 stone (127.00 kg);	none	none
	beer (imperial): 32 gallons (1.455 hl);	36 gallons (1.636 hl)	0.8889
	bran: 6 stone (38.101 kg);	none	none
	malt: 12 stone (76.20 kg);	none	none
	oatmeal: 8 stone (50.802 kg);	none	none
	oats: 14 stone (88.90 kg);	none	none

Unit	Description	English Standard	Deviation
	potatoes: 20 stone (127.00 kg);	none	none
	wine: 6854.4 cu inches (1.123 hl) or 31 1/2 gallons	before 1707: varied with 8883.0 cu inches (ca. 1.46 hl) and 7056.0 cu inches (ca. 1.16 hl) being the most common;	0.7716 0.9714
		after 1707: 7276.5 cu inches (ca. 1.19 hl) or 31 1/2 gallons equal to 1/8 wine tun of 252 gallons	0.9420
Bow	See hoggat	none	none
Bushel	grain: 1740.8 cu inches (28.53 l) or 4 pecks	Winchester: 2150.4 cu inches (35.238 l) or 4 pecks or 8 gallons or 16 pottles or 32 quarts or 64 pints;	0.8095
		imperial: 2219.360 cu inches (36.368 l) or 4 pecks of 8 gallons	0.7844

Appendix 1

Unit	Description	English Standard	Deviation
Crannock	wheat: 8 pecks to 8 bushels (ca. 0.70 to ca. 2.82 hl);	none	none
	oats: 7 to 14 bushels (ca. 2.47 to ca. 4.93 hl)	none	none
Firkin	ale: 2176.0 cu inches (3.566 dkl) or 10 gallons;	8 gallons (ca. 3.70 dkl) equal to 1/2 ale kilderkin or 1/4 ale barrel;	0.9638
	beer (imperial): 8 gallons (3.637 dkl)	9 gallons (4.091 dkl)	0.8889
Gallon	liquids and dry products: 217.6 cu inches (ca. 3.57 l) or 2 pottles or 4 quarts or 8 pints	wine before 1707: varied with 282 cu inches (4.621 l) and 224 cu inches (3.671 l) being the most common;	0.7716/ 0.9714
		wine after 1707: 231 cu inches (3.785 l) or 4 quarts or 8 pints;	0.9420
		ale and beer before 1588: varied considerably;	none
		ale and beer after 1588:	0.7716

Unit	Description	English Standard	Deviation
		282 cu inches (4.621 l) or 4 quarts or 8 pints;	
		grain before 1588: varied between 272 and 282 cu inches (ca. 4.46 to ca. 4.62 l);	0.7993/ 0.7716
		grain after 1588: 268.8 cu inches (4.404 l) or 4 quarts or 8 pints equal to 1/2 peck or 1/8 bushel;	0.8095
		imperial: 277.420 cu inches (4.546 l)	0.7844
Hoggat	grain: 10 bushels (ca. 3.52 hl); synonymous with the bow	none	none
Hogshead	beer (imperial): 52 gallons (2.364 hl)	54 gallons (2.455 hl)	0.9630
Kilderkin	ale: 4352.0 cu inches (7.133 dkl) or 20 gallons or 2 ale firkins;	16 gallons (ca. 7.39 dkl) or 2 ale firkins equal to 1/2 ale barrel;	0.9652
	beer (imperial): 16	18 gallons (8.183 dkl)	0.8889

Unit	Description	English Standard	Deviation
	gallons (7.274 dkl)		
Noggin	liquids: 6.8 cu inches (0.111 l) equal to 1/4 pint	generally 1/2 pint (ca. 0.24 l); sometimes synonymous with the gill	0.4625
Peck	grain: 435.2 cu inches (7.133 l) or 2 gallons	pre-imperial: 537.6 cu inches (8.810 l) or 2 gallons equal to 1/4 Winchester bushel;	0.8095
		imperial: 554.840 cu inches (9.092 l) or 1/4 imperial bushel	0.7844
Pin	beer: 4 gallons (1.818 dkl)	none	none
Pint	liquids and dry products: 27.2 cu inches (0.446 l)	wine: 28.875 cu inches (0.473 l);	0.9420
		ale and beer: 35.25 cu inches (0.578 l);	0.7716
		imperial: 34.677 cu inches (0.568 l) equal to 1/8 gallon	0.7844

Appendix 1

Unit	Description	English Standard	Deviation
Pipe	ale (imperial): 104 gallons (4.728 hl)	108 gallons (4.910 hl)	0.9630
Pottle	liquids: 108.8 cu inches (1.783 l) or 2 quarts	2 quarts (ca. 1.89 l)	0.9434
Puncheon	wine: 84 gallons (ca. 3.00 hl) or 2 tierce	84 gallons (ca. 3.18 hl) or 2 tierce	0.9434
Quart	liquids and dry products: 54.4 cu inches (ca. 0.89 l) or 2 pints	wine: 2 pints (ca. 0.95 l);	0.9368
		ale and beer: 2 pints (ca. 1.16 l);	0.7672
		grain: 2 pints (ca. 1.10 l);	0.8091
		imperial: 69.355 cu inches (1.136 l) or 1/4 gallon	0.7844
Quirren	butter: 4 pounds (1.814 kg) in weight; considered equal to a pottle	none	none
Rundlet	liquids: 3916.8 cu	18 or 18 1/2 gallons	0.9427/

Unit	Description	English Standard	Deviation
	inches (6.420 dkl) or 18 gallons	(ca. 6.81 or ca. 7.00 dkl) equal to 1/14 tun	0.9171
Srone	oatmeal: 12 pounds (5.442 kg) considered equal to 3 pottles	none	none
Tierce	liquids: 9139.2 cu inches (1.498 hl) or 42 gallons	pre-imperial: 42 gallons (ca. 1.59 hl) or 1/2 tertian or 1/3 pipe or 1/6 tun;	0.9421
		imperial: 42 gallons (1.909 hl)	0.7847
Tun	ale: 320 gallons (ca. 11.42 hl) or 8 ale barrels or 16 ale kilderkins or 32 ale firkins	pre-imperial: 252 gallons (ca. 9.54 hl);	1.1971
		imperial: 216 gallons (ca. 9.819 hl)	1.1630
WELSH CAPACITY MEASURES			
Bag	oats in South Wales: 170 quarts (ca. 2.99 hl) or 7 heaped measures or 8 1/2 striked measures	none	none

Appendix 1

Unit	Description	English Standard	Deviation
Barrel	culms: 40 gallons (ca. 1.80 hl) or 4 heaped bushels;	none	none
	lime: 3.25 Winchester bushels (ca. 1.14 hl) or 3 provincial bushels of 10 gallons each	none	none
Bushel	10 gallons (ca. 4.40 dkl) in Brecknothshire;	Winchester and imperial	1.2486/ 1.2098
	10 to 10 1/2 gallons (ca. 4.40 to ca. 4.62 dkl) in Monmouthshire;	same	1.3110/ 1.2703
	20 gallons (ca. 8.80 dkl) in Montgomeryshire	same	2.4972/ 2.4196
Crannock	grain: generally 10 bushels (ca. 3.52 hl)	none	none
Gallon	English standard	See Irish gallon	none
Hobed	lime in South Wales: 4 pedwran of 5 or 6 quarts each (ca. 2.20 to ca. 2.64 dkl)	none	none

Appendix 1

Unit	Description	English Standard	Deviation
	lime in North Wales: 2 storeds (ca. 1.41 hl) or 4 bushels;	none	none
	wheat in North Wales: ca. 173 pounds (ca. 78.471 kg)	English "hobbet" of 168 pounds (76.203 kg)	1.0298
Hoop	grain in Montgomeryshire: 5 gallons (ca. 2.20 dkl)	none	none
Kemple	straw in Midlothian: 40 windlens of 5 to 6 pounds each (90.718 to 108.862 kg)	none	none
Kibin	grain in Anglesey and Carnarvon: 2 pecks (ca. 1.76 dkl) equal to 1/2 bushel	none	none
Llestraid	grain in Cardiff: 20 gallons (ca. 8.81 dkl)	none	none
Measure	oats in South Wales: 20 quarts (0.352 hl)	none	none

Appendix 1

Unit	Description	English Standard	Deviation
Meiliaid	grain in Llandovery: 1/4 bushel (ca. 8.81 l)	none	none
Nive	salt: 7 barrels (ca. 10.36 hl)	none	none
Peccaid	grain in South and East Wales: 5 to 6 gallons (ca. 2.20 to ca. 2.64 dkl); sometimes synonymous with the hobed or hoop	none	none
Pedwran	lime in South Wales: 5 or 6 quarts (ca. 0.55 or ca. 0.66 dkl)	none	none
Quart	English standard	See Irish quart	none
Rhaw	peat: 120 and 140 cu yards (91.747 and 107.038 cu m)	none	none
Sack	wheat in North Wales: 260 pounds (117.933 kg) or 1 1/2 hobeds	none	none
Stack	barley and wheat in	none	none

Unit	Description	English Standard	Deviation
	Glamorganshire: 3 bushels (ca. 1.06 hl); oats in Glamorganshire: 6 bushels (ca. 2.11 hl)	none	none
Stored	grain in North Wales: 2 bushels (ca. 7.05 dkl)	none	none
Tapnet	figs: 20 to 30 pounds (9.072 to 13.608 kg)	none	none
Tub	grain for export in South Wales: 4 bushels (ca. 1.40 hl)	none	none
Windlen	straw in Midlothian: 5 to 6 pounds (2.268 to 2.271 kg)	none	none
IRISH WEIGHTS			
Hundredweight	English standard	generally 112 pounds (50.802 kg) equal to 1/20 ton of 2240 pounds, but many variations ranging from 100 to more than 120 pounds	none

Appendix 1

Unit	Description	English Standard	Deviation
Pound	English standard (also applies to all internal subdivisions)	avoirdupois: 7000 grains (453.592 g) or 256 drams of 27.344 grains each (1.772 g) or 16 ounces of 437 1/2 grains each (28.350 g);	none
		apothecary: 5760 grains (373.242 g) or 288 scruples of 20 grains each (1.296 g) or 96 drams of 60 grains each (3.888 g) or 12 troy ounces of 480 grains each (31.103 g);	none
		mercantile: 6750 troy grains (437.400 g) or 15 mercantile ounces of 450 troy grains each (29.160 g);	none
		troy: 5760 troy grains (373.242 g) or 240 penny-weights of 24 troy grains each (1.555 g) or 12 troy	none

Appendix 1

Unit	Description	English Standard	Deviation
		ounces of 480 troy grains each (31.103 g)	
Stone	English standard	generally 14 pounds (6.350 kg)	none
Ton	English standard	generally 2240 pounds (1016.040 kg) or 20 hundredweight of 112 pounds each, but variations ranged from 2000 to 2400 pounds	none
WELSH WEIGHTS			
Hundredweight	English standard	See Irish hundredweight	none
Maen	wool in South Wales: 26 pounds (12.700 kg) or 4 topstons	none	none
Pound	English standard	See Irish pound	none
Pwn	straw in North Wales: 160 pounds (72.574 kg)	none	none
Pwys	wool in South Wales: 2 pounds (0.907 kg) equal to 1/13 maen	none	none

Appendix 1

Unit	Description	English Standard	Deviation
Stone	English standard	See Irish stone	none
Ton	culm in South Wales: 1904 pounds (863.635 kg) or 17 hundredweight of 112 pounds each;	See Irish ton	0.8500
	lime in North Wales: 1344 pounds (609.625 kg) or 12 hundredweight of 112 pounds each;	same	0.6000
	wax in North Wales: 2688 pounds (1219.250 kg) or 24 hundredweight of 112 pounds each	same	1.2000
Topston	wool in South Wales: 6 1/2 pounds (2.948 kg) equal to 1/4 maen	none	none
Wey	coal in South Wales: 18,144 pounds (8229.937 kg) or 8 tons 2 hundred-weight	none for coal	none

APPENDIX 2
THE QUANTITY MEASURES OF PRE-METRIC EUROPE

The principal quantity measures used in western and central Europe from the late Middle Ages to the nineteenth and twentieth centuries are presented in the following table. Column one contains the name of each unit and its location. Whenever the name of a country is given, it is to be understood that the unit was used in most urban and rural areas within that country. Whenever a unit had local application, the name of a city, county, shire, region, or department is supplied. The designation "British Isles" implies general use throughout England, Scotland, Wales, Ireland, and the Channel Islands; "Italy" includes Sicily, Sardinia, and Malta; and "Germany" extends into eastern Europe to the Slavic regions and south to Austria. In the second column the product uses and variations for each unit are delineated. Column three contains a description of each unit both as to number and, if necessary, physical dimensions. When a certain quantity measure was expressed in terms of length, capacity, volume, or weight, a corresponding metric equivalent is given in parentheses.

Name and Location	Products	Description
Bale[1]		
(England)	almonds	3 hundredweight (146.964 kg)[2]
	bolting cloth	20 pieces[3]
	buckram	60 pieces
	caraway seeds	3 hundredweight (152.406 kg)

Appendix 2

Name and Location	Products	Description
	cochineal	1 1/2 hundredweight (76.203 kg)
	coffee	2 to 2 1/2 hundredweight (101.604 to 127.005 kg)
	cotton yarn	3 to 4 hundredweight (136.077 to 181.436 kg)
	flaxen yarn	240 pounds (108.862 kg)
	fustian	generally 40 or 45 half-pieces
	hay or straw	generally 224 pounds (101.604 kg)
	hemp	20 hundredweight (1016.040 kg)
	licorice	2 hundredweight (101.604 kg)
	paper	10 reams
	pipes	10 gross or 1440 in number
	raw silk	1 to 4 hundredweight (50.802 to 203.208 kg)
	Spanish wool	2 1/4 hundredweight (114.304 kg)
	thread	100 bolts
	wool	180 pounds (81.646 kg)
Balet (England)	piece goods and	generally 1/2 bale

Name and Location	Products	Description
	cloth	
Balla[4]		
(Italy)	piece goods and cloth	any large bundle of goods, usually wrapped in felt or canvas, and tied with rope for transportation by ship or packtrain
Ballen		
(Breslau)	piece goods	10 in number[5]
Balletta		
(Italy)	piece goods and cloth	any small bundle of goods; generally considered equal to 1/2 balla
Ballon		
(France)	colored glass	12 1/2 bundles of 3 plates each
	white glass	25 bundles of 6 plates each
	paper	24 reams, generally of 500 sheets each
Balloncicello		
(Italy)	piece goods and	equivalent to balletta

Appendix 2

Name and Location	Products	Description
	cloth	
Ballone		
(Italy)	piece goods and cloth	equivalent to balla
Ballot		
(France)	piece goods	equivalent to ballon
Bind		
(England)	eels	10 sticks or 250 in number
Binne		
(England)	skins	33 in number
Bolt		
(British Isles)	piece goods and cloth	any bundle or roll
(Berkshire)	osiers	a bundle 42 inches (1.067 m) around and 14 inches (35.56 cm) from the butts
(Essex)	osiers	1/80 load
Bottle		
(England)	hay or straw	a bundle weighing 7 pounds (3.175 kg)
Bouchée		

Name and Location	Products	Description
(Seine-et-Marnais)	onions	a bunch equal to 1/100 grenier
Brawler		
(Somersetshire)	straw	a bundle or sheaf weighing 7 pounds (3.175 kg)
Bunch		
(England)	glass	1/40 waw
	onions or garlic	25 heads
(Cambridgeshire)	osiers	a bundle 45 inches (1.143 m) in circumference at the band
	reeds	a bundle 28 inches (0.711 m) in circumference at the band
(Essex)	teasels	25 heads
(Gloucestershire)	teasels	20 heads for "regulars" and 10 heads for "kings"
(Yorkshire)	teasels	10 heads
Bundle		
(Devonshire)	barley straw	35 pounds (15.876 kg)
(England)	bast ropes	10 in number
	birch brooms	1 or 2 dozen
	brown paper	40 quires
	glovers knives	10 in number

Name and Location	Products	Description
	harness plates	10 in number
(Gloucestershire)	hogshead hoops	36 in number
(Devonshire)	oat straw	40 pounds (18.144 kg)
(Gloucestershire)	osiers	1 1/4 feet (0.457 m) in circumference
(Hampshire)	osiers	42 inches (1.067 m) around the lower band
(Worcestershire)	osiers	38 inches (0.965 m) in circumference
(Yorkshire)	straw for thatching	1/12 thrave
(Devonshire)	wheat straw	28 pounds (12.700 kg)
(Hamborough)	yarn	20 skein
Cage		
(England)	quails	generally 28 dozen
	other animals	no fixed number
Carrat[6]		
(Sicily)	pipe staves	4 squares of 5 1/12 palmi in length or approximately 3800 staves
Cent		
(British Isles)	piece goods	equivalent to hundred

Name and Location	Products	Description
(France)	most goods	100 in number
	timber	100 solives or 300 cu Parisian pieds (10.283 st)
(Nantes)	small Norwegian boards	124 in number
(River Sèvre)	salt	28 muids (699 hl)
(Charente, Loire, Vendée)	salt	100 setiers (366 hl)
Centaine		
(France)	piece goods	100 in number
Cord[7]		
(England)	wood	a double cube of 4 feet or 128 cu feet (3.624 cu m)
(Derbyshire)	wood	128, 155, and 162 1/2 cu feet (3.624, 4.389, and 4.601 cu m)
(Gloucestershire)	wood	approximately 78 cu feet (2.209 cu m)
(Sussex)	wood	126 cu feet (3.568 cu m)
Decher or Dechent		
(Germany)	piece goods	10 in number
Dicker		

Appendix 2

Name and Location	Products	Description
(England and	most goods	10 in number
Scotland)	horseshoes and gloves	10 pairs
	necklaces	10 bundles of 10 necklaces each
Douzaine		
(France)	piece goods	12 in number; rarely 13 or 14
Dozen		
(British Isles)	piece goods	equivalent to douzaine
Dozzina		
(Italy)	piece goods	equivalent to douzaine
Dutzend		
(Germany)	piece goods	equivalent to douzaine
Fad		
(England)	straw	a bundle equal to 1/12 thrave
Fadge		
(England)	sticks	any bundle or bale
Fagotto		
(Italy)	cloth	any small bundle or bale
Fangot		
(England)	raw silk	a bundle weighing 1 to 2 3/4 hundredweight (50.802 to

Name and Location	Products	Description
		139.705 kg)
	yarn (grogram and mohair)	a bundle weighing 1 1/2 to 2 1/2 hundredweight (76.203 to 127.005 kg)
Fardel		
(England)	generally cloth	any large bale or bundle
Fardlet		
(England)	generally cloth	any small bale or bundle
Fargot		
(Department of Nord)	piece goods and cloth	any bale or bundle varying in weight from 150 to 160 livres[8]
Fascicule[9]		
(France)	medical and culinary goods	a handful of ordinary ingredients
Fatt		
(England)	bristles	5 hundredweight (254.010 kg)
	coal	1/4 chalder (ca. 3.17 hl)
	isinglass	3 1/4 to 4 hundredweight (147.417 to 181.436 kg)
	unbound books	4 bales equal to 1/2 maund
	wire	20 to 25 hundredweight

Name and Location	Products	Description
		(1016.040 to 1270.050 kg)
	yarn	220 or 221 bundles
Fesse[10]		
(England)	hay or straw	any bale larger than a bottle but smaller than a truss
Flitch		
(England)	cured hog meat	a side
Flock		
(England)	piece goods	40 in number or sets
Gerbe		
(France)	cut wheat	any bale or bundle
Glean		
(England)	herrings	1/15 rees or 25 in number
(Essex and Gloucestershire)	teasels	1 bunch
Globen		
(Germany)	flax	15 ends
Grenier		
(France)	onions	100 in number
Gross		
(British Isles and	piece goods	small of 12 dozen or 144 in

Name and Location	Products	Description
Germany)		number; large of 12 small gross or 1728 in number[11]
Gwyde		
(England)	eels	10 sticks or 250 in number
Hank		
(England)	yarn (cotton or spun silk)	7 skeins or 840 yards (7.681 hm)
	yarn (worsted)	7 wraps or 12 cuts or 560 yards (5.121 hm)
Hasp		
(England)	linen yarn	6 heers or 1/4 spindle or 3600 yards (3291.840 m)
Hattock		
(Northern England)	grain	10 or 12 sheaves; similar to a shock[12]
Heap		
(Scotland)	limestone	4 1/4 cu yards (3.249 cu m) and weighing approximately 5 tons
Heer		
(England)	linen yarn	2 leas or 1/6 hasp or 600 yards (548.640 m)

Appendix 2

Name and Location	Products	Description
Hundert		
(Germany)	piece goods	small generally of 100 in number; large generally of 120 in number
(Danzig)	piece goods	small 120 in number; large of 12 Ring or 48 Schock or 2880 in number
Hundred		
(British Isles)	most goods	100 in number
	balks, barlings, boards, canvas, capravens, cattle, deals, eggs, faggots, herrings, lambskins, linen cloth, nails, oars, pins, poles, reeds, spars, staves, stockfish, stones, tile, and wainscoats	120 in number
	cod, ling, salt-fish, and haberdine	124 in number

Name and Location	Products	Description
	"hardfish"	160 in number
	onions and garlic	225 in number
(Roxburghshire and Selkirkshire)	lambs and sheep	106 in number
(Fifeshire)	herrings	132 in number
Jointée		
(France)	grain	any amount capable of being scooped up with both hands; a double handful
Kiepe		
(Lübeck)	piece goods	600 in number
Kip		
(England)	goat skins	50 in number
	lamb skins	30 in number
Knot[13]		
(Essex)	wool yarn	80 turns around a reel
Knitch		
(Northern England and Scotland)	unbroken straw	a bundle or sheaf with a circumference of 34 inches (8.636 dm)
Last		

Name and Location	Products	Description
(England)	herrings	20 mease or 10,000 to 12,600 in number
Laste		
(Lille)	wheat	38 rasières (26.668 hl)
(Marseille)	import and export items	3 quintaux or 300 mercantile livres (12.238 dcg)
(Montpellier)		2 milliers or 20 quintaux or 2000 livres (ca. 2000 kg)
Lest		
(France)	herrings	12 caques or barils (3.524 hl)
Load		
(Essex)	osiers	80 bolts of no standard dimensions
Loggin		
(Yorkshire)	straw	a bundle weighing 14 pounds (6.350 kg)
Mandel		
(Germany)	piece goods	15 in number
Manipule		
(France)	medical and pharmaceutical	a handful of ordinary ingredients

Appendix 2

Name and Location	Products	Description
Marque		
(Normandy)	timber	300 chevilles of 12 cu pouces each or 2 1/12 cu Parisian pieds in all (0.2285 dst)
Maund		
(England)	unbound books	2 fatts or 8 bales or 40 reams
Mazzo		
(Italy)	wool and yarn	any bunch, bale, or bundle
Mease		
(England)	herrings	500 to 630 in number equal to 1/20 last
Membrure		
(France)	firewood	any stack or pile
Migliaio		
(Italy)	piece goods	1000 in number
Mil		
(British Isles)	piece goods	equivalent to thousand
Nest		
(England)	piece goods	3 in number or sets
Paar		
(Germany)	piece goods	2 in number or sets

Appendix 2

Name and Location	Products	Description
(Breslau)	fox fur	12 skins
Pack		
(England)	cloth	generally 10 pieces
	flax and flour	240 pounds (108.862 kg)
	teasels	generally 9000 heads for "kings" and 20,000 heads for "middlings"
(Gloucestershire)		40 staffs or 1000 gleans or 20,000 heads for "middlings" and 30 staffs or 900 gleans or 9000 heads for "kings"
(Yorkshire)		1350 bunches of 10 heads each or 13,500 in all
(Huddersfield)	vegetables	240 pounds (108.862 kg)
(Yorkshire and Lancashire)	lambs wool	44 pounds (19.958 kg)
(Brunswick and Leipzig)	cloth	10 pieces
Packet		
(England)	piece goods	any small pack or bundle
Pair		

Name and Location	Products	Description
(British Isles)	piece goods	2 in number or sets
Pezzo		
(Italy)	cloth	any small bolt
Piling		
(Staffordshire)	wheat straw	a bundle consisting of 3 sheaves
Pincée		
(France)	medical and culinary goods	a pinch or spoonful of any substance
Poignée		
(France)	medical and pharmaceutical	a handful of ordinary ingredients
Pugille		
(France)	medical and pharmaceutical	a pinch or spoonful of any substance
Pwn		
(North Wales)	straw	a pack or burden weighing 160 pounds (72.574 kg)
Quarteron		
(France)	piece goods	one-fourth of any group of items sold by number

Appendix 2

Name and Location	Products	Description
Quire		
(British Isles)	paper	24 or 25 sheets equal to 1/20 ream[14]
Ream		
(British Isles)	paper	"regular" of 20 quires of 24 or 25 sheets each equal to 1/10 bale "printers" of 21 1/2 quires or 516 sheets "stationers" of 504 sheets
Rees		
(England)	herrings	15 gleans or 375 in number
Riem		
(Bremen)	paper	2 Ries
Ries		
(Bremen)	paper	1/2 Riem
Ring		
(Danzig)	piece goods	240 in number or 2 small hundreds of 120 each
(Hamburg)	most goods	1000 in number
	lumber	240 boards

Name and Location	Products	Description
Roll[15]		
(England)	parchment	60 skins
Rook		
(Yorkshire)	beans	4 sheaves set up to dry in a field
Rope		
(England)	onions and garlic	15 heads or 1/15 hundred of 225[16]
Roul		
(England)	eels	1500 in number
Ruck		
(Derbyshire)	bark	5 1/4 stacked cu yards (4.014 cu m)
Sack		
(Scotland)	sheepskins	500 in number
Saum		
(Breslau)	cloth	22 pieces
(Leipzig)	cloth	10 pieces
Schifflast		
(Hamburg)	oxhoft staves	1800 in number
	pipe staves	1200 in number

Appendix 2

Name and Location	Products	Description
	tun staves	2400 in number
Schock		
(Germany)	piece goods	4 Mandel or 60 in number
Score		
(British Isles)	most goods	20 in number
(Liverpool)	barley, beans, and oats	21 bushels (ca. 7.40 hl)
(Newcastle)	coal	21 chalders (124,656 pounds or 56,542.71 kg)
(Roxburghshire and Selkirkshire)	grain	21 bolls (ca. 58.59 hl)
(Derbyshire)	lime	20 to 22 heaped bushels (ca. 9.01 to ca. 9.91 hl)
(Dumbartonshire)	sheep	21 in number
Sechzig		
(Danzig)	wagon parts	60 large Hundert or 7200 in all
Sheaf		
(England)	grain	1/12 to 1/24 thrave
	steel	30 gads or pieces of uncertain weight equal to 1/6 or 1/12

Name and Location	Products	Description
		burden
Shock		
(England)	piece goods	60 in number
(Derbyshire)	corn	12 sheaves
Skive		
(Southampton)	teasels	generally 500 in number
Staff		
(Essex)	teasels	1250 in number or 50 gleans of 25 teasels each
(Gloucestershire)		500 in number or 25 gleans of 20 teasels each for "middlings" and 300 or 30 gleans of 10 teasels each for "kings"
Stick		
(England)	eels	25 in number equal to 1/10 bind or gwyde
Stiege		
(Germany)	piece goods	20 in number; sometimes 30
Stoke[17]		
(England)	dinnerware	60 pieces
Tausend		

Appendix 2

Name and Location	Products	Description
(Germany)	piece goods	small generally of 1000 in number; large generally of 1200 in number
Telleron		
(France)	wood	any stack or pile
Thousand		
(British Isles)	piece goods	generally 10 times larger than the corresponding hundred
Thrave		
(England)	straw	12 to 24 sheaves
Timber[18]		
(England)	certain fur skins	40 in number
Tonne		
(Danzig and Königsberg)	herrings	13 Wall or 1040 in number
Torsa		
(Italy)	piece goods and cloth	any large bundle or bale
Torsello		
(Italy)	mainly cloth	any long, flat bale used for

Name and Location	Products	Description
		transporting goods by animals
Truss		
(England)	hay or straw	a large bundle commonly regarded as weighing 56 pounds (25.401 kg)
Trussell		
(England)	skins and cloth	any bundle or bale capable of being carried
Wall		
(Germany)	piece goods	80 in number
Warp		
(Sussex and Kent)	herrings	4 in number
Waw		
(England)	glass	40 bunches of uncertain weight
Web		
(Scotland)	window glass	60 bunches of uncertain weight
Webe		
(Germany)	piece goods	72 in number
Wrap		
(England)	worsted yarn	80 yards (73.152 m) or 1/7 hank of 560 yards

Appendix 2

Name and Location	Products	Description
Zehnling		
(Bremen)	skins	10 in number
Zimmer		
(Germany)	piece goods	4 Decher or 40 in number

APPENDIX 3
THE WEIGHTS OF PRE-METRIC EUROPE

The principal weights and weight systems used in western and central Europe from the late Middle Ages to the late nineteenth century are presented in the following table. "France" is used here to refer to the present national boundaries and, with few exceptions, does not extend to the medieval parameters that included Belgium, Flanders, and Brabant. "Italy," however, is interpreted broadly to include Sicily, Sardinia, and Malta. "Germany" extends into eastern Europe to Bohemia and south to Austria. The emphasis throughout is placed on the major urban and rural centers of commerce, business, finance, industry, and agriculture.

All units have separate listings in column one, together with the location at which they were employed. Whenever a nation is named as the principal site, it is to be understood that the unit was either a statutory or government promulgated standard. In every other instance the name of a city, province, region, or department is supplied; this generally denotes a local or sectional standard or variant. The second column explains how each standard or variant was employed. The last column contains a description of each unit.

Name and Location	Application	Description
Achtel (Vienna)	gold	2 Pfennigen (2.188 g)
Acino[1]		

Appendix 3

Name and Location	Application	Description
(Naples)	gold, silver, and other precious materials	1/20 trappeso (4.455 cg) equal to 1/600 oncia or 1/7200 libbra
Argento		
(Cagliari)	precious metals and jewels	36 grani (1.694 g)
As		
(Baden, Karlsruhe)	precious metals and jewels	(0.050 g)
Balance[2]		
(Champagne)	dry products	(ca. 1300 - ca. 1500): of undetermined size
Breton[3]		
(Bretagne)	generally dry products	(ca. 1400 - ca. 1800): 100 livres of any product
Cantaio		
(Italy)	all products	(ca. 1600 - ca. 1800): a hundredweight equivalent to cantaro
Cantarello[4]		
(Sardinia)	all products	(ca. 1600 - ca. 1800): 104

Name and Location	Application	Description
		libbre (42.28 kg) or 4 rubbi of 26 libbre each
Cantaro[5]		
(Ancona)	all products unless indicated otherwise	light of 100 libbre (32.96 kg) or 4 rubbi of 25 libbre each
		heavy of 150 libbre (49.44 kg) or 6 rubbi of 25 libbre each
(Cagliari)		commercial of 100 libbre (40.66 kg) or 4 rubbi of 25 libbre each
		heavy of 104 libbre (42.28 kg)
(Cantania)		275 libbre (79.34 kg) or 100 rotoli of 30 once each
(Ferrara)		100 libbre (34.51 kg)
(Florence)		(after 1836): legal of 100 libbre (33.95 kg)[6]
	wool and fish	160 libbre (54.33 kg)
(Genoa)		150 libbre (47.65 kg) or 6 rubbi or 100 rotoli equal to 1/5 peso
		heavy of 150 libbre (52.30 kg)

Name and Location	Application	Description
		or 6 rubbi or 100 rotoli
(Livorno)		same as Florence
	sugar	151 libbre (51.27 kg)
	wool and fish	same as Florence
(Malta)		250 libbre (79.38 kg) or 100 rotoli of 30 once each
		heavy of 275 libbre (87.07 kg) or 100 rotoli
(Messina)		250 libbre (79.39 kg) or 100 rotoli
		heavy of 275 libbre (87.33 kg) or 100 rotoli
(Milan)		(after 1803): 100 libbre (100.00 kg) or 10 rubbi[7]
(Naples)		100 libbre (32.08 kg) or 36 rotoli
		light of 150 libbre (48.11 kg)
		heavy of 277 3/4 libbre (89.10 kg) or 100 rotoli
(Palermo)		same as Messina
(Rome)		light of 100 libbre (33.91 kg)

Name and Location	Application	Description
		or 10 decine equal to 1/10 migliaio
		heavy of 1000 libbre (339.29 kg) or 10 light cantari or 100 decine
(Treviso)		100 libbre (51.67 kg)
(Venice)		light of 100 libbre (30.12 kg)
		heavy of 100 libbre (47.70 kg)
(Vicenza)		100 libbre (48.65 kg)
Carara[8]		
(Livorno)	all products	(ca. 1600 - ca. 1800): a local name for the cantaro
Carat[9]		
(France)	precious metals and jewels	4 grains valued at 3.876 grains poids de marc (2.059 dg) and equal to 1/144 once (for pearls and diamonds) or 1/1152 marc[10]
Carato	precious metals and jewels unless indicated otherwise	all local or regional standards consisted of 4 grains:[11]
(Albenga)		(1.833 dg)

Name and Location	Application	Description
(Bologna)		(1.885 dg) equal to 1/10 ferlino or 1/160 oncia
(Genoa)		same as Albenga
(Ferrara)		(1.800 dg)
(Florence)		(1.965 dg) equal to 1/144 oncia
(Livorno)		same as Florence
(Milan)		(2.057 dg)
(Modena)		(1.770 dg)
	silk	same as Bologna
(Naples)		(2.056 dg) or 32 sedicesimi equal to 1/130 oncia
(Rome)		(1.962 dg)
(Turin)		(2.135 dg) equal to 1/1152 marca
(Venice)		(2.070 dg) equal to 1/144 oncia or 1/1152 marca
Carica		
(Venice)	all products	400 libbre light weight (120.51 kg) or 4 light cantari
Carro		

Name and Location	Application	Description
(Brescia)	all products	100 pesi (802.03 kg)
(Crema)	all products	100 light pesi (759.44 kg)
(Modena)	all products	100 pesi (851.14 kg)
Centesimo		
(Naples)	generally valuable commodities	(after 1840): 10 trappesi (0.0089 kg) (before 1840): frequently the hundredth part of any weight or measure[12]
Centimillesimo		
(Naples)	generally valuable commodities	(after 1840): (0.0000089 kg) (before 1840): sometimes employed in the sciences as the hundred-thousandth part of any weight or measure
Centinaio		
(Italy)	all products	a hundredweight equivalent to cantaro
Charge[13]		
(France)	all products unless indicated	300 livres poids de marc (146.85 kg) or 3 quintaux

Name and Location	Application	Description
	otherwise	
(Marseille)	grain	300 livres poids de marc (122.38 kg) or 4 émines or 12 civadiers
(Nantes)		300 livres (148.32 kg)
(Nice)		300 livres (93.48 kg)
(Toulouse)		3 quintaux (127.26 kg)[14]
Coccio		
(Sicily)	gold and silver	8 ottavi (0.055 g)
Colpo		
(Sardinia)	lime	1040 libbre (422.82 kg) or 10 cantarelli
Coppia[15]		
(Florence and Lucca)	silk	2 libbre (0.678 kg)
Danapeso		
(Italy)	precious metals and jewels	a pennyweight equivalent to denaro
Decimillesimo		
(Naples)	generally valuable commodities	(after 1840): (0.000089 kg)

Name and Location	Application	Description
Decimo		
(Naples)	generally valuable commodities	(after 1840): 100 trappesi (89.100 g)
Decina		
(Naples)	wool	11 1/9 libbre (3.564 kg) or 4 rotoli
(Rome)	dry products	10 libbre (3.391 kg) equal to 1/10 light cantaro or 1/100 heavy cantaro[16]
Degalatro		
(Northern Italy)	all products	4 rotoli of any substance; sometimes synonymous with decimo or decina
Dekas		
(Baden, Karlsruhe)	gold	(0.500 g)
Demi-carat		
(France)	precious metals and jewels	2 grains equal to 1/2 carat valued at 1.938 grains poids de marc (1.0295 dg)
Demi-gros		
(France)	generally valuable	(before 1812): 1/2 gros

Name and Location	Application	Description
	commodities	(1.912 g) (1812-40): (1.95 g)[17]
Demi-livre (France)	all products	2 quarterons, 4 demi-quarterons, 8 onces, or 16 demi-onces equal to 1/2 standard or regional livre[18]
Demi-once (France)	generally valuable commodities	1/2 standard or regional once[19]
Demi-quarteron (France)	generally valuable commodities	2 onces or 4 demi-onces equal to 1/2 standard or regional quarteron[20]
Demi-quintal (France)	all products	1/2 standard or regional quintal
Demi-scrupule (France)	medical and pharmaceutical	1/2 standard or regional scrupule
Denaro	precious metals	all local or regional

Name and Location	Application	Description
	and jewels unless indicated otherwise	standards consisted of 24 grani[21]
(Acqui)		(1.129 g)
(Albenga)		(1.100 g)
(Ancona)		(1.144 g)
(Ascoli Piceno)		(1.177 g)
(Bergamo)		(2.258 g)
(Bobbio)		same as Albenga
(Brescia)		(1.114 g)
(Carrara)		(1.128 g)
(Casale Monferrato)		same as Acqui
(Como)		(1.099 g)
(Crema)		(1.130 g)
(Cremona)		(1.075 g)
(Domodossola)		(1.135 g)
(Florence)		(2.432 g)
	gold	(1.179 g)
(Genoa)		same as Albenga
(Guastalla)		(1.127 g)
(Lodi)		same as Brescia
(Lucca)		(1.161 g)

Name and Location	Application	Description
	gold	same as Florence
(Massa)		(1.145 g)
(Milan)		(after 1803): (1.000 g)
		(before 1803): same as Domodossola
	gold	(1.224 g)
(Modena)		same as Guastalla
(Mortara)		(1.107 g)
(Novara)		same as Crema
(Pallanza)		same as Domodossola
(Parma)		(1.139 g)
(Pavia)		same as Mortara
(Pesaro)		same as Ancona
(Perugia)		(1.173 g)
(Piacenza)		(1.103 g)
(Pistoia)		(1.123 g)
(Porto Maurizio)		same as Albenga
(Rome)		same as Ascoli Piceno
(Senigallia)		(1.163 g)
(Tortona)		same as Crema
(Turin)	gold	(1.281 g)

Name and Location	Application	Description
(Varallo)		(1.225 g)
(Venice)	gold	(1.242 g)
(Voghera)		(1.109 g)
Denier		
(France)	generally gold, silver, precious metals, and jewels	(before 1800): 24 grains (1.275 g) equal to 1/3 gros, 1/24 once, or 1/192 marc; occasionally considered synonymous with the scrupule[22] (1800-12): 10 new grains (1 g) equal to 1/10 new gros of 10 g or 1/100 new once of 100 g[23]
Dimarc		
(France)	all products	equivalent to the livre poids de marc of 16 onces; hence two (=di) marcs of 8 onces each[24]
Doppelzentner		
(Germany)	all products	(100 kg) before metric usage
Drachme[25]		
(France)	medical and	60 grains (3.824 g) or 3

Appendix 3

Name and Location	Application	Description
	pharmaceutical	scrupules or 6 oboles equal to 1/8 once
Drachme		
(Aachen)	medical and pharmaceutical at all sites	(3.649 g)
(Berlin)		(before 1816): 3 Scrupeln (3.720 g)
		(after 1816): 3 Scrupeln (3.654 g)
(Darmstadt)		3 Scrupeln (3.727 g)
(Frankfurt)		1/128 Pfund:
		light (3.691 g)
		heavy (3.835 g)
(Hamburg)		(after 1856): 3 Scrupeln (3.720 g)
(Karlsruhe)		3 Scrupeln (3.906 g)
(Kassel)		3 Scrupeln (3.726 g)
(Vienna)		(4.387 g)
Dramma	medical and pharmaceutical	all local or regional standards consisted of 72

Appendix 3

Name and Location	Application	Description
	at all sites	grani or 3 denari or scrupoli equal to 1/8 oncia:
(Brescia)		(1.671 g) or 4 quarte
(Cagliari)		(3.202 g)
(Cesena)		same as Bologna
(Cremona)		(3.224 g)
(Ferrara)		(3.595 g)
(Florence)		(3.537 g)
(Genoa)		(3.299 g)
(Lucca)		(3.484 g)
(Massa)		(3.435 g)
(Milan)		(3.404 g)
(Modena)		(3.546 g)
(Naples)		(2.673 g)
(Palermo)		(3.306 g)
(Parma)		(3.417 g)
(Pistoia)		(3.370 g)
(Rome)		(3.532 g)
(Turin)		same as Cagliari
(Urbino)		(3.391 g)
(Venice)		(3.138 g)

Name and Location	Application	Description
Dukaten		
(Cologne)	monetary	1/67 Mark (3.490 g)
(Vienna)	monetary	60 Gran (3.520 g)
Engel		
(Cologne)	monetary	1/152 Mark (1.538 g)
Esche		
(Cologne)	gold and silver	1/4352 Mark (0.0537 g)
Estelin		
(France)	gold and silver	28.8 grains (1.53 g) or 2 oboles or mailles or 4 félins equal to 1/20 once or 1/160 marc[26]
Fascio	dry products unless indicated otherwise	all local or regional standards consisted of 100 heavy libbre:
(Como)		(79.16 kg)
(Lodi)		(74.84 kg)
(Milan)		(76.25 kg)
(Mortara)		(74.37 kg)
(Novara)		(75.94 kg)
(Pallanza)		(87.14 kg)

Name and Location	Application	Description
(Pavia)		same as Mortara
(Varallo)	hay	(84.34 kg)
(Voghera)	lime and gypsum	(74.52 kg)
Félin		
(France)	gold and silver	7 1/5 grains (0.38 g) equal to 1/2 maille, 1/4 estelin, 1/10 gros, 1/80 once, or 1/640 marc
Ferlino	gold, silver, and other precious materials at all sites	all local or regional standards consisted of 10 carati generally equal to 1/16 oncia or 1/192 libbra:
(Bologna)		(1.885 g)
(Ferrara)		(1.797 g)
(Modena)		(1.773 g)
Frachtspfund (Bremen)	shipping goods	(149.56 kg)
Galatro		
(Apulia)	cotton	1/25 of any local or regional cantaro
Glied		

Name and Location	Application	Description
(Fulda)	wool	1/5 Wollzentner (10.709 kg)
Grain		
(France)	all products	(before 1800): 24 primes (53.115 mg) equal to 1/12 félin, 1/24 denier, 1/72 gros, 1/576 once, 1/4608 marc, or 1/9216 livre[27]
		(1800-12): (0.1 g)
		(1812-40): (5.425 cg)
Gran		
(Augsburg)	gold and silver	(0.82 g)
(Berlin)	medical	(before 1816): (0.0620 g)
		(after 1816): (0.0609 g)
(Cologne)	gold and silver	1/288 Mark (0.82 g)
(Darmstadt)	medical	(0.0621 g)
(Hamburg)	medical	(after 1856): (0.0620 g)
	gold and silver	(0.0510 g)
(Karlsruhe)	gold and silver	4 Granchen (0.06 g)
	medical	(0.0651 g)
(Kassel)	medical	(0.0621 g)
(Vienna)	gold and silver	1/60 Dukaten (0.0587 g)

Name and Location	Application	Description
Granchen		
(Karlsruhe)	gold and silver	4 Richttheilen (0.015 g)
Grano	all products unless indicated otherwise	all local or regional standards consisted of 24 granottini or granotti equal to 1/24 denaro:
(Acqui)		(0.000047 kg)[28]
(Albenga)		(0.000046 kg)[29]
(Ancona)		(0.000048 kg)[30]
(Ascoli Piceno)		(0.000049 kg)[31]
(Cagliari)	medical	(0.000053 kg)
(Cremona)		(0.000045 kg)
(Ferrara)		same as Cremona
	medical	(0.000050 kg)
(Milan)	gold, diamonds, and jewels	(0.000051 kg) (after 1803): (0.000100 kg)
(Naples)		(0.000051 kg)
(Palermo)	medical	8 ottavi (0.000055 kg)
(Turin)		(0.000053 kg)
(Venice)		(0.000052 kg)
Granottino		

Name and Location	Application	Description
(Piedmont)	generally valuable commodities	1/24 grano (ca. 0.002 g) or 1/576 denaro
Granotto		
(Piedmont)	generally valuable commodities	equivalent to granottino
Gros[32]		
(France)	all products	(before 1800): 72 grains (3.824 g) or 2 1/2 estelins, 3 deniers or scrupules, 5 mailles, or 10 félins equal to 1/8 once, 1/64 marc, or 1/128 livre (1800-12): 100 new grains (10 g) equal to 1/10 new once of 100 g (1812-40): 72 grains usuels (3.906 g) equal to 1/128 livre usuelle of 500 g[33]
Grosso		
(Northern Italy)	medical and pharmaceutical	(before 1803): equivalent to dramma

Name and Location	Application	Description
		(after 1803): 100 grani (10 g) or 10 denari equal to 1/100 libbra nuova italiana[34]
Handelspfund		
(Hildesheim)	commercial goods	0.99844 Hanoverian or Prussian Pfund (0.467 kg)
Heller		
(Berlin, Bremen, Brunswick, and Darmstadt)	gold and silver at all sites	(1816-72): 128 Richtpfennigen (0.457 g)
(Kassel)		256 Richtpfennigen (0.457 g)
(Leipzig)		(after 1837): 1/1024 Pfund (0.457 g)
(Cologne)		1/912 Mark (0.256 g)
(Frankfurt)		1/1024 Pfund (0.494 g)
Hüttenzentner		
(Germany)	mining	heavy Zentner (10 to 20 Pfund in excess)
Karat		
(Austria)	precious gems unless indicated	(0.208 g)

Name and Location	Application	Description
	otherwise	
(Berlin)		1/2, 1/4, 1/8, 1/16, 1/32, and 1/64 divisions (0.205 g)
	gold and silver	(10.62 g)
(Cologne)	gold and silver	(10.62 g)
(Darmstadt)		(0.206 g)
(Frankfurt)		same as Austria
(Hamburg)		4 Gran (0.204 g)
(Karlsruhe)		4 Gran (0.240 g)
(Kassel)		same as Darmstadt
(Leipzig)	gold and silver	same as Cologne
Karch		
(Austria)	wholesale shipments	(224.01 kg)
Kleuder		
(Hanau)	wool	1/5 Zentner (10.516 kg)
(Kassel)	wool	21 heavy Pfund (10.169 kg)
Kommerzlast		
(Bremen)	commercial shipping at all sites	(after 1858): 6000 Pfund (3000 kg)
(Emden)		6000 Pfund (2981.11 kg)
(Hamburg)		(before 1858): 1 1/4

Appendix 3

Name and Location	Application	Description
		Schiffslast (2423.05 kg)
		(after 1858): 6000 Pfund
		(3000 kg)
Korn		
(Altenburg, Detmold)	precious gems	(after 1858): 1/10 Zent (0.0167 g)
Krämerpfund		
(Bremen)	small commercial items	(0.4703 kg)
Lägel		
(Austria)	steel	1/2 Saum (70.015 kg)
Läp		
(Breslau)	commercial goods	24 Pfund (9.733 kg)
Last		
(Berlin)	bulk-rated shipments	(after 1816): 4000 Pfund (1870.59 kg)
(Bremen)	salt	(1818-58): 4000 Pfund (1994.00 kg)
	coal	(after 1858): 12,000 Pfund (6000 kg)
	grain crops	(after 1860): 2600 to 4800

Name and Location	Application	Description
		Pfund (1300 to 2400 kg)
(Brunswick)	butter	large of 280 Pfund (130.96 kg)
		small of 224 Pfund (104.76 kg)
(Königsberg)	stone	1980 Pfund (927.70 kg)
(Vienna)	bulk-rated shipments	(2240.05 kg)
Libbra	all products unless indicated otherwise	(ca. 1600 - ca. 1800): all local or regional standards consisted of 12 once unless indicated otherwise:[35]
(Abruzzo e Molise)		(0.321 kg)
(Basilicata)		same as Abruzzo e Molise
(Calabria)		same as Abruzzo e Molise
(Campania)		same as Abruzzo e Molise
(Emilia-Romagna)	jewels	8 once (0.239 kg): Cesena
		(0.317 kg): Bobbio, Piacenza
		(0.324 kg): Reggio nell´ Emilia, Guastalla
	medical	(0.326 kg): Bologna, Cesena
		(0.328 kg): Parma
		(0.330 kg): Cesena

Name and Location	Application	Description
	gold	(0.339 kg): Cesena
		(0.340 kg): Modena
		(0.345 kg): Ferrara, Rimini
		(0.348 kg): Ravenna
		(0.362 kg): Bologna, Modena
		(for gold)
		middle of 18 once (0,518 kg): Rimini
		heavy of 24 once (0.691 kg): Rimini
(Friuli-Venezia Giulia)		(0.301 kg): Udine
		heavy (0.477 kg): Udine
(Lazio)		(0.339 kg): Rome, Civitavecchia, Viterbo
(Liguria)	medical	(0.307 kg): Porto Maurizio, S. Remo
		light (0.317 kg): Genoa, Albenga, La Spezia, Porto Maurizio, S. Remo
		heavy (0.318 kg): Genoa, La Spezia

Name and Location	Application	Description
(Lombardy)		(0.309 kg): Cremona (0.319 kg): Mortara, Pavia, Voghera (0.321 kg): Brescia, Lodi (0.325 kg): Bergamo, Crema (0.327 kg): Como, Milan heavy of 28 once (0.744 kg): Pavia, Mortara heavy of 28 once (0.745 kg): Voghera heavy of 28 once (0.748 kg): Lodi middle of 28 once (0.759 kg): Crema heavy of 28 once (0.762 kg) and 32 once (0.871 kg): Milan heavy of 30 once (0.813 kg): Bergamo heavy of 30 once (0.814 kg): Crema (after 1803): 10,000 grani

Name and Location	Application	Description
		(1.000 kg) or 10 once, 100 grossi, or 1000 denari: Milan
(Malta)		(0.317 kg)
		commercial of 30 once (0.791 kg)
(Marche)		(0.325 kg): Urbino
		(0.330 kg): Ancona, Pesaro
		(0.335 kg): Senigallia
		small balance (0.339 kg): Ancona, Ascoli Piceno
		large steelyard (0.353 kg): Ascoli Piceno
		middle of 18 once (0.494 kg): Pesaro
		heavy of 24 once (0.659 kg): Pesaro
(Piedmont)	medical	(0.307 kg) throughout
		(0.317 kg): Novi Ligure
		heavy (0.318 kg): Novi Ligure
		(0.325 kg): Acqui, Casale Monferrato, Novara

Name and Location	Application	Description
		(0.326 kg): Tortona
		(0.327 kg): Domodossola
		mark weight (0.352 kg): Novara
		(0.369 kg): Turin
		14 once (0.381 kg): Domodossola
		16 once (0.436 kg): Domodossola
		heavy of 28 once (0.759 kg): Novara
		heavy Milanese of 28 once (0.762 kg): Domodossola, Pallanza, Novara
		marine of 30 once (0.814 kg): Novara
		heavy of 32 once (0.871 kg): Domodossola, Pallanza
		heavy of 36 once (0.980 kg): Domodossola
(Puglia)		(0.321 kg)
(Sardinia)	medical	(0.307 kg)

Name and Location	Application	Description
	gold	(0.325 kg)
		commercial (0.407 kg)
(Sicily)		(0.317 kg)
(Tuscany)		(0.323 kg): Pistoia
		(0.325 kg): Carrara
		(0.330 kg): Massa
		(0.334 kg): Lucca
		(0.339 kg): Arezzo, Lucca, Florence, Pistoia, Livorno, Pisa, Siena
		regional standard (0.341 kg): Lucca
(Umbria)		(0.338 kg): Perugia, Spoleto
(Venezia Euganea)		(0.301 kg): Venice
	hay	(0.321 kg): Vicenza
		(0.333 kg): Verona
		(0.339 kg): Padua, Treviso, Vicenza
		heavy (0.477 kg): Rovigo, Venice
		heavy (0.486 kg): Padua

Name and Location	Application	Description
		heavy (0.500 kg): Verona
		heavy (0.517 kg): Treviso
Libbretta		
(Italy)	all products	any light libbra
Liespfund		
(Berlin)	small shipments	(before 1816): 14 Pfund
	at all sites	(6.546 kg)
		(after 1816): 14 Pfund
		(6.548 kg)
(Bremen)		(1818-58): 14 Pfund (6.979 kg)
(Dantzig)		16 1/2 Pfund (7.717 kg)
(Hamburg)		14 Pfund (6.784 kg)
(Hannover)		(before 1835): 14 Pfund
		(6.854 kg)
(Königsberg)		16 1/2 Pfund (7.731 kg)
Lira		
(Italy)	all products	equivalent to libbra
Liretta		
(Italy)	all products	(ca. 1700 - ca. 1800):
		equivalent to libbretta
Livre[36]		

Appendix 3

Name and Location	Application	Description
(France)	all products unless indicated otherwise	(ca. 800 - ca. 1350): "livre d´esterlin" or "livre de Charlemagne" of 5760 grains (367.1 g) or 20 sous, 12 onces, 240 deniers, or 480 oboles[37]
		(ca. 1350 - ca. 1800): "livre poids de marc" or "livre de Paris" of 9216 grains (489.506 g)[38]
	medical	(ca. 1350 - 1732): 5760 grains (367.1 g) or 12 onces, 96 drachmes, 288 scrupules, or 576 oboles[39]
	silk	mercantile of 15 onces (459 g)
		(1800-12): metric livre of 1 kg
		(1812-40): "livre usuelle" of 500 g or 1/2 kg or 2 demi-livres, 4 quarterons, 16 onces, or 128 gros
(Abbeville)		(422 g)

Appendix 3

Name and Location	Application	Description
(Aire)		(428.3 g)
(Aix-en-Provence)		mercantile or mint (379.1 g)
(Albi)		(407.9 g)
(Allevard)		mining (532.5 g)
		industrial (552.6 g)
(Apt)		(397.5 g)
(Arles)		(391.2 g)
(Aude)		(469 g)
(Aveyren)		(408 g)
(Avignon)		(408.7 g)
(Beaucaire)		(412.9 g)
(Bourges)		(468.3 g)
(Cambrai)		(470 g)
(Carpentras)		(400 g)
(Castres)		(412 g)
(Clermont)		same as Aire
(Douai)		(425 g)
(Dunkirk)		same as Aire
(Embrun)		mercantile or mint (435 g)
(Gap)		mercantile or mint (392 g)
(Grenoble)		mercantile or mint (417.3 g)

Name and Location	Application	Description
		urban (442.7 g)
		Savoie (551.8 g)
		industrial (552.6 g)
(Istres)		(388.5 g)
(La Rochelle)		(404.4 g)
(Lavaur)		same as Albi
(Lille)		(431.3 g)
(Limoges)		(481.9 g)
(Marseille)		same as Albi
(Mauberge)		(467.1 g)
(Moissac)		(429.1 g)
(Montauban)		mercantile or mint (425.6 g)
(Montpellier)		same as Albi
(Morlaix)		(491.3 g)
(Nancy)		(455.7 g)
(Nantes)		(494.4 g)
(Nîmes)		same as Albi
(Rambereillers)		(460.1 g)
(Rouen)		agricultural (509.1 g)
		wool (528.6 g)
(Salon)		(376.6 g)

Name and Location	Application	Description
(St. Chamas)		(379.2 g)
(St. Mitre)		same as Salon
(Tarascon)		(388.1 g)
(Toulon)		(406.5 g)
(Toulouse)		mercantile or mint (413.6 g)
		meat (1240.8 g)
(Tours)		(475.7 g)
(Troyes)		(520 g)
(Uzès)		(412.1 g)
(Vienne)		balance (404.1 g)
		hooked (456.8 g)
(Villefranche)		(436.8 g)
(Voiron)		mercantile or mint (417.3 g)
Loth		
(Aachen)	medical unless indicated otherwise	4 Drachmen (0.0146 kg)
(Altenburg)		(after 1858): 1/30 Pfund (16.667 g)
(Augsburg)		1/16 Mark (14.754 g)
(Bamberg)		1/32 Pfund (14.637 g)
(Berlin)		4 Quentchen (0.0144 kg)

Appendix 3

Name and Location	Application	Description
	gold and silver	(after 1816): 4 Quentchen (14.616 g)
(Bremen)		(1818-58): 4 Quentchen (0.0156 kg)
		(after 1858): Neuloth (0.0500 kg)
	gold and silver	4 Quentchen (14.616 g)
(Breslau)		4 Quentchen (0.0127 kg)
(Brunswick)		(before 1858): 4 Quentchen (0.0146 kg)
(Coblenz)		1/32 Pfund (14.5732 g)
(Danzig)		4 Quentchen (0.01462 kg)
(Darmstadt)		(before 1821): 4 Quentchen (0.01462 kg)
		(1821-72): (0.01562 kg)
	gold and silver	4 Quentchen (14.6212 g)
(Detmold)		(before 1858): 4 Quentchen (0.01461 kg)
		(after 1858): 10 Quentchen (0.01667 kg)
(Dresden)		1/32 Pfund (14.6004 g)

Name and Location	Application	Description
(Frankfurt)		light (14.7632 g)
		heavy (15.3398 g)
(Fulda)		1/32 Pfund (15.9366 g)
(Hamburg)		4 Quentchen (0.0152 kg)
		(after 1858): Neuloth (0.0500 kg)
	gold and silver	(0.0146 kg)
(Hannover)		(before 1835): 4 Quentchen (0.0152 kg)
		(1835-58): (0.0144 kg)
		(after 1858): (0.0500 kg)
(Karlsruhe)		4 Quentchen (0.0156 kg)
(Kassel)		light (0.0146 kg)
		heavy (0.0151 kg)
(Leipzig)		1/16 Mark (14.6133 g)
(Zollverein)		(after 1840): 10 Quentchen (16.667 g)
Maille		
(France)	gold and silver	(ca. 800 - ca. 1350): "maille d´esterlin" of 14.4 grains (ca. 0.8 g) or 2 félins equal

Name and Location	Application	Description
		to 1/2 estelin, 1/40 once, 1/320 marc, or 1/640 livre[40] "maille de denier" of 12 grains (ca. 0.7 g) equal to 1/2 denier of 24 grains[41]
Marc		
(Paris)	principally gold and silver	4608 grains (244.753 g) or 8 onces, 64 gros, 160 estelins, 192 deniers, 320 mailles, or 640 félins equal to 1/2 livre poids de marc
(France)	all products at all sites	(1812-40): 1/2 "livre usuelle" of 250 g
(Limoges)		(240.93 g)
(Montpellier)		(239.12 g)
(Tours)		(237.87 g)
(Troyes)		(260.05 g)[42]
Marca		
(Italy, Sicily, and Sardinia)	principally gold and silver at all sites	8 once or 192 denari equal to 1/2 heavy libbra

Name and Location	Application	Description
(Asti)[43]		(0.246 kg)
(Genoa)		(0.211 kg)
(Valsesia)[44]		(0.235 kg)
(Venice)		(0.238 kg)[45]
Mark	gold and silver at all sites	all 8 Unzen:
(Aachen)		(0.2336 kg)
(Augsburg)		(0.2361 kg)
(Berlin)		(after 1816): (0.2338 kg)
(Bonn)		(0,2336 kg)
(Bremen)		(before 1818): (0.2338 kg)
		(1818-58): (0.2492 kg)
(Breslau)		(0.2032 kg)
(Cologne)		(0.2338 kg)
(Danzig)		(0.2335 kg)
(Darmstadt)		(0.2339 kg)
(Frankfurt)		light (0.2362 kg)
		heavy (0.2454 kg)
(Hamburg)		(0.2339 kg)
(Hannover)		(before 1835): (0.2448 kg)
		(after 1835): (0.2336 kg)

Name and Location	Application	Description
(Karlsruhe)		(0.2500 kg)
(Kassel)		(before 1861): (0.2339 kg) (1861-71): (0.2500 kg)
(Königsberg)		(0.1959 kg)
(Leipzig)		(0.2338 kg)
(Nuremberg)		(0.2378 kg)
(Vienna)		(0.2808 kg)
Migliaio[46]		
(Italy)	all products	1000 libbre or 10 light cantari
Millesimo		
(Naples)	generally valuable commodities	(after 1840): synonymous with a trappeso of 0.00089 kg equal to 1/10 centesimo (before 1840): sometimes the thousandth part of any weight or measure
Millier		
(France)	all products	1000 livres poids de marc (489.51 kg) or 10 quintaux equal to 1/2 marine tonneau[47]

Name and Location	Application	Description
Moggione		
(Voghera)	all products	(ca. 1500 - 1800): (149.04 kg)
Obole		
(France)	gold and silver	(ca. 800 - ca. 1350): "obole d´esterlin" of 14.4 grains (ca. 0.8 g) or 2 félins equal to 1/2 estelin, 1/40 once, 1/320 marc, or 1/640 livre "obole de denier" of 12 grains (ca. 0.7 g) equal to 1/2 denier of 24 grains[48]
Obolo		
(Naples)	gold and silver	10 acini (0.445 g)
(Rome)	gold and silver	12 grani (0.588 g) or 3 silique equal to 1/2 scrupulo, 1/6 dramma, or 1/48 oncia
Octavo		
(Italy)	generally medical and pharmaceutical	(ca. 1600 - ca. 1800): equivalent to ottavo
Once		
(France)	all products	(ca. 800 - ca. 1350): "once

Name and Location	Application	Description
	unless indicated otherwise	d´esterlin" of 480 grains (30.59 g) or 20 deniers or 40 oboles equal to 1/12 livre de Charlemagne
		"once poids de marc" of 576 grains (30.594 g) or 2 demi-onces, 8 gros or drachmes, 24 scrupules or deniers, or 13,824 primes equal to 1/2 demi-quarteron, 1/8 marc, or 1/16 livre
	gold, silver, pearls, and diamonds	576 grains (30.594 g)[49]
		(1800-12): metric once of 100 g or 10 gros of 10 g each, 100 deniers of 1 g each, or 1000 grains of 0.1 g each
		(1812-40): "once usuelle" of 31.250 g or 8 gros or 576 new grains equal to 1/4 quarteron, 1/8 marc, or 1/16 livre

Appendix 3

Name and Location	Application	Description
Oncia	all products unless indicated otherwise	(ca. 1600 - ca. 1800): all local or regional standards were equal to 1/12 libbra:
(Acqui)		8 ottavi (0.0271 kg)
(Albenga)		8 ottavi or dramme (0.0264 kg)
(Ancona)		8 ottavi or dramme (0.0275 kg)
(Ascoli Piceno)		8 ottavi (0.0283 kg)
(Bergamo)		12 denari (0.0271 kg)
(Bobbio)		8 ottavi (0.0264 kg)
(Bologna)		8 ottavi or 16 ferlini (0.0301 kg)
	medical	8 dramme (0.0271 kg)
(Brescia)		16 dramme (0.0267 kg)
	medical	8 dramme (0.0267 kg)
(Cagliari)		4 quarte (0.0339 kg)
	medical	8 dramme (0.0256 kg)
(Carrara)		24 denari (0.0271 kg)
(Casale Monferrato)		8 ottavi (0.0271 kg)
(Cesena)		8 ottavi (0.0275 kg)
	medical	8 dramme (0.0271 kg)
	gold	(0.0283 kg)

Appendix 3

Name and Location	Application	Description
	jewels	4 quarts, 24 denari, or 144 carati (0.0298 kg)
(Como)		24 denari (0.0264 kg)
(Crema)		24 denari (0.0271 kg)
(Cremona)		24 denari (0.0258 kg)
(Domodossola)		8 ottavi (0.0273 kg)
(Ferrara)		4 quarte (0.0288 kg)
	medical	8 dramme (0.0288 kg)
(Florence)		24 denari (0.0283 kg)
	medical	8 dramme (0.0283 kg)
(Genoa)		8 ottavi (0.0264 kg)
	medical	8 dramme (0.0264 kg)
	gold	4 quarte (0.0264 kg)
(Guastalla)		24 denari (0.0270 kg)
(Lodi)		24 denari (0.0267 kg)
(Lucca)		24 denari (0.0279 kg)
	medical	8 dramme (0.0279 kg)
	gold	24 denari (0.0283 kg)
(Malta)		16 parti or 32 trappesi (0.0264 kg)
(Massa)		24 denari (0.0275 kg)

Appendix 3

Name and Location	Application	Description
	medical	8 dramme (0.0275 kg)
(Messina)		30 trappesi (0.0267 kg)
(Milan)		24 denari (0.0272 kg)
	medical	8 dramme (0.0272 kg)
	gold	24 denari (0.0299 kg)
		(after 1803): 1000 grani (0.1000 kg) or 10 grossi or 100 denari equal to 1/10 libbra nuova italiana
(Modena)		16 ferlini (0.0284 kg)
	medical	8 dramme (0.0284 kg)
(Mortara)		24 denari (0.0266 kg)
(Naples)		30 trappesi (0.0267 kg)
	medical	10 dramme (0.0267 kg)
	diamonds	130 carati (0.0267 kg)
(Padua)		6 saggi (0.0282 kg)
		heavy of 6 saggi (0.0405 kg)
(Palermo)		4 quarte (0.0264 kg)
	medical	8 dramme (0.0264 kg)
	gold	30 trappesi, 600 acini, or 9600 sedicesimi (0.0264 kg)

Name and Location	Application	Description
(Pallanza)		8 ottavi or 24 denari (0.0272 kg)
(Parma)		24 denari (0.0273 kg)
	medical	8 dramme (0.0273 kg)
(Pavia)		24 denari (0.0266 kg)
(Perugia)		8 ottavi (0.0281 kg)
(Piacenza)		24 denari (0.0265 kg)
(Pistoia)		8 dramme (0.0270 kg)
(Porto Maurizio)		8 ottavi (0.0264 kg)
(Ravenna)		8 ottavi (0.0290 kg)
(Rimini)		8 ottavi (0.0284 kg)
(Rome)		8 ottavi (0.0283 kg)
	medical	8 dramme (0.0283 kg)
	gold	24 denari (0.0283 kg)
(Senigallia)		8 ottavi (0.0279 kg)
(Tortona)		24 denari (0.0271 kg)
(Treviso)		light (0.0282 kg)
		heavy (0.0431 kg)
(Turin)		8 ottavi (0.0307 kg)
	medical	8 dramme (0.0256 kg)
(Urbino)		8 dramme (0.0271 kg)

Name and Location	Application	Description
(Varallo)		24 denari (0.0301 kg)
(Venice)		light (0.0251 kg)
		heavy (0.0397 kg)
	medical	8 dramme (0.0251 kg)
(Verona)		light (0.0278 kg)
		heavy (0.0417 kg)
(Vicenza)		light (0.0282 kg)
		heavy (0.0405 kg)
(Voghera)		8 ottavi (0.0266 kg)
Ort		
(Bremen)	gold and silver	(1818-58): (0.974 g)
Ortchen		
(Hannover)	gold and silver	(after 1835): (0.0009 kg)
Ottavo	medical and pharmaceutical unless indicated otherwise	all local or regional standards were generally 3 denari, equal to 1/8 oncia, and synonymous with the dramma:[50]
(Acqui)		(0.003389 kg)
(Albenga)		(0.003299 kg)
(Ancona)		(0.003433 kg)

Name and Location	Application	Description
(Ascoli Piceno)		(0.003532 kg)
(Bobbio)		same as Albenga
(Bologna)		2 ferlini (0.003769 kg)
(Cagliari)		2 sedicesimi (0.004235 kg)
(Casale Monferrato)		same as Acqui
(Cesena)		(0.003435 kg)
(Domodossola)		(0.003404 kg)
(Ferrara)		2 ferlini (0.003595 kg)
(Genoa)		light (0.003299 kg)
		heavy (0.003309 kg)
(Malta)		72 grani (0.003307 kg)
(Modena)	silk	20 carati (0.003769 kg)
(Pallanza)		same as Domodossola
(Perugia)		(0.003519 kg)
(Pesaro)		same as Ancona
(Porto Maurizio)		same as Albenga
(Ravenna)		(0.003623 kg)
(Rimini)		(0.003599 kg)
(Rome)		same as Ascoli Piceno
(Senigallia)		(0.003488 kg)
(Turin)		(0.003842 kg)

Name and Location	Application	Description
(Voghera)		(0.003327 kg)
Parto		
(Malta)	generally valuable commodities	(1.649 g)
Pellet[51]		
(France)	weight of account	1/24 prime (0.0922 mg) or 1/576 grain
Perrée		
(France)	all products	generally 8 livres poids de marc (3.916 kg)[52]
Pesata		
(Cagliari)	firewood	150 libbre (60.98 kg)
(Genoa)	firewood	5 cantari (238.25 kg)
Peso		
(Bergamo)	all products unless indicated otherwise	10 heavy libbre (8.150 kg)
(Bologna)		25 libbre (9.046 kg)
(Brescia)		25 libbre (8.020 kg)
(Crema)		10 light libbre (7.594 kg) 10 heavy libbre (8.137 kg)

Name and Location	Application	Description
(Cremona)		25 libbre (7.737 kg)
	flax	27 libbre (8.356 kg)
(Genoa)		750 libbre (237.72 kg) or 5 cantari, 30 rubbi, or 500 rotoli
		heavy (261.50 kg)
(Malta)		5 rotoli (3.969 kg) equal to 1/20 cantaro
(Milan)		10 heavy libbre (7.625 kg)
(Modena)		25 libbre (8.511 kg)
(Naples)	lime	40 rotoli (35.64 kg) or 111 1/9 libbre
(Parma)		25 libbre (8.200 kg)
(Piacenza)		25 libbre (7.938 kg)
(Rome)	hay	300 libbre (101.70 kg)
	lime and gypsum	400 libbre (135.63 kg)
(Verona)		25 light libbre (8.332 kg)
Peson		
(France)	all products	any small weight, usually varying from 2 to 4 onces
Pfennig		

Appendix 3

Name and Location	Application	Description
(Berlin)	gold and silver at all sites	(after 1816): 2 Heller (0.913 g)
(Bremen)		same as Berlin
(Breslau)		(0.791 g)
(Brunswick)		(0.913 g)
(Cologne)		same as Brunswick
(Darmstadt)		2 Heller (0.914 g)
(Frankfurt)		(0.986 g)
(Hamburg)		(0.946 g)
(Karlsruhe)		4 Karat (0.977 g)
(Kassel)		2 Heller (0.914 g)
(Lübeck)		(0.953 g)
(Vienna)		(1.094 g)
Pfund		
(Aachen)	all products unless indicated otherwise	2 Mark (0.467 kg)
(Altenburg)		(after 1858): 30 Loth (500 g)
(Anhalt)	medical	(349.832 g)
(Arolsen)		light (467.41 g)

Appendix 3

Name and Location	Application	Description
		heavy (476.35 g)
		(after 1858): (500 g)
(Augsburg)		light (0.4727 kg)
		heavy (0.4912 kg)
(Bamberg)		32 Loth (468.384 g)
(Berlin)		32 Loth (0.468 kg)
	medical	(before 1816): 12 Unzen (0.3576 kg)
		(after 1816): (0.3508 kg)
(Bremen)		(1818-58): 2 Mark (0.4984 kg)
		(after 1858): 10 Neuloth (0.500 kg)
(Breslau)		32 Loth (0.4053 kg)
(Brunswick)		(before 1858): 32 Loth (0.4677 kg)
(Coblenz)		32 Loth (0.4663 kg)
(Cologne)		2 Mark (467.6246 g)
(Danzig)		32 Loth (0.4677 kg)
(Darmstadt)		light (0.4679 kg)
		heavy (0.5053 kg)
		(after 1821): (0.500 kg)

Appendix 3

Name and Location	Application	Description
(Detmold)		(before 1858): 32 Loth (0.4674 kg)
		(after 1858): (0.500 kg)
(Dillenburg)		(0.4414 kg)
(Dresden)		(0.4669 kg)
(Düsseldorf)		(467.6246 g)
(Elberfeld)		same as Düsseldorf
(Elbing)		(424.9512 g)
(Emden)		light (468.536 g)
		heavy (496.851 g)
(Erfurt)		(472.060 g)
(Frankfurt)		light (0.4724 kg)
		heavy (0.4919 kg)
(Fulda)		(509.97 g)
(Gotha)		(after 1858): (500 g)
(Hamburg)		32 Loth (0.4846 kg)
	banking	2 Mark (0.4677 kg)
(Hannover)		(before 1835): 2 Mark (0.4896 kg)
		(after 1835): (0.4672 kg)
		(after 1858): (500 g)

Name and Location	Application	Description
(Heidelburg)		light (467.970 g)
		heavy (505.408 g)
(Holstein)		(483.3998 g)
(Karlsruhe)		10 Zehnlinge (0.500 kg)
(Kassel)		light (0.4678 kg)
		heavy (0.4842 kg)
		(1861-71): (0.500 kg)
(Kiel)		(476.5311 g)
(Kleve)		(0.4672 kg)
(Königsberg)		(468.536 g)
(Leipzig)		32 Loth (467.6246 g)
(Lübeck)		32 Loth (486.47 g)
(Mannheim)		(494.9340 g)
(Mecklenburg)		(484.726 g)
(Munich)		(560.8993 g)
(Munster)		(475.8831 g)
(Nuremberg)	commercial	(509.996 g)
	silver	(477.138 g)
	medical	(357.854 g)
(Rostock)	city	(508.229 g)
	commercial	(484.028 g)

Name and Location	Application	Description
(Salzburg)		(559.9921 g)
(Siegen)		(0.4677 kg)
(Strassburg)		(470.8288 g)
(Ulm)		(468.7552 g)
(Vienna)		32 Loth (0.5602 kg)
(Württenberg)		(467.8352 g)
(Würzburg)		(477.0495 g)
(Zollverein)		(after 1840): 30 Loth (0.500 kg)
Pipe		
(Hamburg)	oil	820 Pfund (397.3797 kg)
Prime		
(France)	medical and pharmaceutical	1/24 grain (2.213 mg)
Quarantottesimo		
(Florence)	gold and silver	(0.001023 g)
Quarro		
(Venice)	precious metals and jewels	equivalent to dramma
Quarta	gold and silver at all sites	all local or regional standards equal to 1/4 oncia:

Name and Location	Application	Description
(Brescia)		(0.000418 kg)
(Cagliari)		2 ottavi (0.008470 kg)
(Ferrara)		2 ottavi (0.007190 kg)
(Genoa)		36 carati (0.000660 kg)
(Malta)		2 ottavi (0.006615 kg)
(Palermo)		2 dramme (0.006612 kg)
(Venice)		6 denari or 36 carati (0.000745 kg)
Quarteron		
(France)	all products at all sites	(before 1812): 4 onces (122.38 g) or 2 demi-quarterons equal to 1/4 livre poids de marc or 1/2 marc or demi-livre (1812-40): 4 onces (125 g) or 1/4 livre usuelle traditionally: 1/4 cent or centaine (here signifying 1/4 hundredweight)[53]
(Toulouse)		mercantile or mint of 26 livres (10.75 kg)

Appendix 3

Name and Location	Application	Description
Quarterone		
(Italy)	all products	1/4 of any cantaro
Quartiere		
(Italy)	all products	1/4 of any libbra
Quent		
(Altenburg)	gold and silver at all sites	(1.667 g)
(Austria)		(4.3750 g)
(Berlin)		(0.0036 kg)
(Bremen)		(1818-58): (0.0039 kg)
		(after 1858): (0.0050 kg)
(Breslau)		(0.003164 kg)
(Brunswick)		(before 1858): (0.003654 kg)
(Coblenz)		(3.6433 g)
(Cologne)		(3.6533 g)
(Danzig)		(0.003654 kg)
(Darmstadt)		(before 1821): (0.003655 kg)
		(after 1821): (0.003908 kg)
(Detmold)		(before 1858): (0.003652 kg)
		(after 1858): (0.001667 kg)
(Dresden)		(3.6501 g)

Name and Location	Application	Description
(Frankfurt)		(3.9430 g)
(Fulda)		(3.9841 g)
(Hamburg)		(before 1858): (0.0038 kg)
		(after 1858): (0.0050 kg)
(Hannover)		(before 1835): (0.0038 kg)
		(after 1835): (0.0036 kg)
		(after 1858): (0.0050 kg)
(Karlsruhe)		(0.003908 kg)
(Kassel)		light (0.003655 kg)
		heavy (0.003783 kg)
(Leipzig)		(3.6533 g)
(Lübeck)		(3.7876 g)
(Zollverein)		(after 1840): (1.667 g)
Quentlein		
(Leipzig)		4 Pfennigen (3.6533 g)
Quintal		
(France)	all products at all sites	a hundredweight of 100 livres
(Paris)		"poids de marc" of 100 livres (48.951 kg) equal to 1/3 charge or 1/10 millier

Name and Location	Application	Description
		(after 1840): metric of 100 kg[54]
(Toulouse)[55]		
Quintale		
(Italy)	all products	a hundredweight equivalent to cantaro
Richtspfennig		
(various sites)	gold and silver	generally (0.0036 g)
Richttheile		
(Karlsruhe)	gold and silver	1/4 Granchen (0.00000375 kg)
Robe		
(France)	all products	(ca. 1600 - 1750): 26 livres
Roggenlast		
(Emden)	commercial goods	4000 Pfund (1987.404 kg)
(Hannover)	commercial goods	(after 1835): 4000 Pfund (1868.8 kg)
		(after 1858): (2000 kg)
Rotolo	all products at all sites	all local or regional standards generally equal to 1/100 cantaro:[56]
(Aquila)		33 1/3 once (0.891 kg) or

Name and Location	Application	Description
		2 7/9 libbre or 1000 trappesi[57]
(Emilia-Romagna)		18 once (0.475 kg) or 1 1/2 libbre
(Liguria)		same as Emilia-Romagna
(Malta)		legal of 30 once (0.793 kg) or 2 1/2 libbre
		heavy of 33 once (0.872 kg) or 2 3/4 libbre
(Sicily)		same as Malta
(Tuscany)		36 once (1.019 kg) or 3 libbre
Rubbio	principally for oil at all sites	all local or regional standards consisted generally of 25 libbre equal to 1/4 cantaro:[58]
(Ancona)		(8.240 kg)
(Bergamo)		(8.128 kg)
(Biella)		(before 1818): (9.221 kg)
		(after 1818): (9.222 kg)[59]
(Cagliari)		(10.16 kg)
(Cerro Tenaro)		(8.134 kg)[60]
(Como)		(7.916 kg)

Name and Location	Application	Description
(Lodi)		(8.018 kg)
(Milan)		(8.170 kg)[61] (after 1803): 10 libbre (10.00 kg), 100 once, 1000 grossi, 10,000 denari, or 100,000 grani
(Naples)		(8.340 kg)
(Novara)		(8.137 kg)
(Novi Ligure)		light (7.919 kg) heavy (7.942 kg)[62]
(Tortona)		(8.141 kg)
(Valenza)		(7.968 kg)[63]
(Varallo)		(9.037 kg)
(Voghera)		(7.984 kg)
Rup		
(France)	all products	(ca. 1600 - ca. 1750): 25 livres
Saggio		
(Venezia Euganea and Liguria)	principally for spices, precious gems, gold, and	light of 96 grani (4.185 g) or 24 carati equal to 1/6 oncia

Name and Location	Application	Description
	silver	heavy of 128 grani (6.626 g) or 32 carati equal to 1/6 oncia
Salma		
(Cagliari)	salt	(569.19 kg)
Salmata		
(Varallo)	all products	12 rubbi (108.44 kg)
Sarcinée		
(Toulouse)	goods shipped by land transport	3 quintaux (ca. 127 kg); used interchangeably with charge
Saum		
(Austria)	steel	(154.0336 kg)
Schiffslast		
(Austria)	commercial shipments at all sites	(1120.024 kg)
(Berlin)		(after 1816): 4000 Pfund (1870.844 kg)
(Bremen)		(1818-58): 4000 Pfund (1993.600 kg) (after 1858): (2000 kg)
(Brunswick)		4000 Pfund (1870.8 kg)

Name and Location	Application	Description
(Hamburg)		6000 Pfund (2907.657 kg)
		(after 1858): (2000 kg)
(Hannover)		(after 1835): 4000 Pfund (1868.8 kg)
		(after 1858): (2000 kg)
Schiffspfund		
(Aachen)	commercial shipments at all sites	318 Pfund (148.520 kg)
(Berlin)		330 Pfund (154.345 kg)
(Bremen)		(1818-58): 308 Pfund (153.507 kg)
(Breslau)		396 Pfund (160.593 kg)
(Brunswick)		280 Pfund (130.956 kg)
(Danzig)		330 Pfund (154.341 kg)
(Dresden)		330 Pfund (154.181 kg)
(Emden)		300 Pfund (149.055 kg)
(Hamburg)		280 Pfund (135.691 kg) and 320 Pfund (155.075 kg)
(Hannover)		(before 1835): 280 Pfund (137.098 kg)
(Königsberg)		330 Pfund (154.617 kg)

Name and Location	Application	Description
(Leipzig)		330 Pfund (154.316 kg)
Scrupolo	precious metals, jewels, medical and pharmaceutical at all sites	all local or regional standards consisted generally of 24 grani equal to 1/3 dramma or 1/24 oncia; synonymous with denaro:
(Bologna)		(1.131 g)
(Cagliari)		20 grani (1.067 g)
(Cesena)		same as Bologna
(Cremona)		(1.075 g)
(Ferrara)		(1.198 g)
(Florence)		(1.179 g)
(Genoa)		(1.100 g)
(Lucca)		(1.161 g)
(Massa)		(1.145 g)
(Milan)		(1.135 g)
(Modena)		(1.182 g)
(Naples)		2 oboli (0.891 g)
(Palermo)		20 cocci (1.102 g)
(Parma)		(1.139 g)
(Rome)		(1.177 g)

Name and Location	Application	Description
(Turin)		same as Cagliari
(Urbino)		(1.130 g)
(Venice)		light of 20 grani (1.046 g)
Scrupule		
(France)	precious metals, jewels, medical and pharmaceutical	(before 1812): 24 grains (1.275 g) equal to 1/3 gros or drachme, 1/24 once, or 1/384 livre (1812-40): (1.302 g)
Scudo		
(Turin)	gold and jewels	(3.344 g)
Sechszehntal		
(Vienna)	gold and jewels	1/256 Mark (1.1 g)
Sedicesimo		
(Cagliari)	precious gems	(ca. 1700 - ca. 1800): (0.002117 kg)
(Naples)		(ca. 1700 - ca. 1800): (0.006425 kg)
Sedicino		
(Genoa)	silk	(ca. 1700 - ca. 1800): (0.001650 kg)

Appendix 3

Name and Location	Application	Description
Siliqua		
(Rome)	precious metals and jewels	4 grani (0.000196 kg) equal to 1/3 obolo or 1/6 denaro
(Venice)	precious metals and jewels	(0.000208 kg); synonymous with the carato
Sizain		
(France)	generally valuable commodities	144 grains (7.648 g) or 2 gros or 6 deniers equal to 1/4 once
Skrupel		
(Berlin)	medical at all sites	(before 1816): 20 Gran (0.001240 kg) (after 1816): (1.217997 g)
(Darmstadt)		20 Gran (1.24246 g)
(Hamburg)		(after 1856): 20 Gran (0.001240 kg)
(Kassel)		20 Gran (1.24188 g)
(Nuremberg)		(1.331 g)
(Vienna)		(1.458 g)
Sou		
(France)	generally gold, silver, and other	(ca. 800 - ca. 1350): 288 grains (18.356 g) or 12 deniers

Name and Location	Application	Description
	valuable items	or 24 oboles equal to 1/20 livre de Charlemagne (after 1350): 192 grains (12.238 g) or 12 deniers or 24 oboles equal to 1/20 mint marc or 2/3 sou d´esterlin
Stalln		
(Dillenburg)	iron	160 Pfund (70.626 kg)
(Siegen)	iron	170 Pfund (79.512 kg)
Stein		
(Baden)	all products at all sites	1/10 Zentner (5.001 kg)
(Berlin)		(after 1816): 22 Pfund (10.290 kg)
(Bremen)		(1818-58): 20 Pfund (9.968 kg)
(Breslau)		22 and 24 Pfund (8.917 and 9.727 kg)
(Brunswick)		10 Pfund (4.677 kg)
(Danzig)		22 and 33 Pfund (10.289 and 15.434 kg)

Name and Location	Application	Description
(Dresden)		22 Pfund (10.279 kg)
(Frankfurt)		22 Pfund (10.393 kg)
(Hamburg)		light of 10 Pfund (4.846 kg)
		heavy of 20 Pfund (9.692 kg)
(Hannover)		10 Pfund (4.896 kg)
(Karlsruhe)		10 Pfund (5.000 kg)
(Kassel)		22 Pfund (10.292 kg)
(Königsberg)		light of 20 Pfund (9.371 kg)
		heavy of 33 Pfund (15.462 kg)
(Leipzig)		22 Pfund (10.288 kg)
(Vienna)		(11.202 kg)
Tonne		
(Bremen)	oil	216 Pfund (107.654 kg)
	liquids	(1818-58): 2000 Pfund (996.800 kg)
		(after 1858): 2000 Pfund (1000 kg)
(Hamburg)	butter	small of 224 Pfund (108.552 kg)
		large of 280 Pfund (135.691 kg)

Name and Location	Application	Description
	liquids	(after 1858): 2000 Pfund (1000 kg)
(Hannover)	liquids	(after 1835): 2000 Pfund (934.4 kg)
(Leipzig)	oil	224 Pfund (104.748 kg)
Tonneau[64]		
(France)	all products	marine of 2000 livres (979.112 kg) or 2 milliers or 20 quintaux[65] (after 1795): metric of 10 quintaux (1000 kg)[66]
Tonnellata		
(Ancona)	bulkrated ship goods at all sites	3000 libbre (988.75 kg) or 20 cantari
(Florence)		2000 libbre (679.08 kg)
(Milan)		(after 1803): 1000 libbre (1000 kg)
(Naples)		1140 rotoli (1015.74 kg)
Trappeso		
(Caltanissetta)	gold, silver, and gems at all sites	16 cocci (0.8816 g) equal to 1/30 oncia or 1/360 libbra[67]

Name and Location	Application	Description
(Malta)		18 grani (0.8245 g) equal to 1/32 oncia or 1/384 libbra
(Messina)		20 cocci (0.8910 g) equal to 1/30 oncia or 1/360 libbra
(Naples)		20 acini (0.8910 g) equal to 1/30 oncia or 1/360 libbra; considered synonymous with scrupolo
Treseau		
(France)	pharmaceutical	3 deniers (3.875 g) equal to 1/8 once; synonymous with gros
Uchau		
(Toulouse)	valuable items	1/8 once (4.308 g)[68]
Unze		
(Austria and Germany)	all products	generally 2 Loth or 8 Drachmen
Vague		
(France)	unknown	of undetermined size, origin, and location[69]
Ventiquattresimo		

Name and Location	Application	Description
(Rome)	gold and precious gems	(ca. 1700 - ca. 1800): (0.002044 g)
Vierling		
(Karlsruhe)	small items at all sites	4 Unzen or 8 Loth (0.1248 kg)
(Nuremberg)		(127.6 g)
(Vienna)		1/4 Pfund or 8 Loth (140 g)
Waage		
(Bremen)	iron at all sites	(1818-58): 120 Pfund (59.808 kg)
(Dresden)		44 Pfund (20.557 kg)
(Leipzig)		44 Pfund (20.575 kg)
(Nassau)		(55.99 kg)
Wollzentner		
(Fulda)	wool	5 Glied (53.547 kg)
Zehnling		
(Germany)	small products	(50 g)
Zent		
(Germany)	precious metals	(after 1840): (0.01667 g)
Zentas		
(Germany)	precious metals	10 Dekas or 100 As (0.005 kg)

Name and Location	Application	Description
Zentner		
(Aachen)	commercial goods at all sites	106 Pfund (49.506 kg)
(Arolsen)		light (50.480 kg) heavy (51.446 kg)
(Baden)		100 Pfund (50 kg)
(Berlin)		110 Pfund (51.447 kg)
(Bremen)		(1818-58): 116 Pfund (57.814 kg) (after 1858): 100 Pfund (50 kg)
(Breslau)		132 Pfund (53.531 kg)
(Cleve)		110 Pfund (51.375 kg)
(Coblenz)		100 Pfund (46.634 kg)
(Cologne)		106 Pfund (49.568 kg)
(Danzig)		110 Pfund (51.447 kg)
(Darmstadt)		(after 1821): 100 Pfund (50 kg)
(Detmold)		(before 1858): 108 Pfund (50.486 kg) (after 1858): 100 Pfund

Name and Location	Application	Description
		(50 kg)
(Dresden)		(51.393 kg)
(Düsseldorf)		110 Pfund (51.439 kg)
(Emden)		100 Pfund (49.685 kg)
(Fulda)		100 Pfund (50.997 kg)
(Hamburg)		112 Pfund (54.276 kg)
		(after 1858): 100 Pfund (50 kg)
(Hannover)		(before 1835): 112 Pfund (54.839 kg)
		(after 1835): 100 Pfund (46.72 kg)
		(after 1858): 100 Pfund (50 kg)
(Heidelberg)		108 Pfund (50.541 kg)
(Karlsruhe)		100 Pfund (50 kg)
(Kassel)		(1861-71): 100 Pfund (50 kg)
(Königsberg)		110 Pfund (51.539 kg)
(Leipzig)		110 Pfund (51.439 kg)
(Vienna)		100 Pfund (56.046 kg)
(Zollverein)		(after 1840): (50 kg)

Name and Location	Application	Description
Zethim		
(Germany)	monetary	1/2 Loth (0.0076 kg)
Zollloth		
(Germany)	gold and silver	1/30 Zollpfund (0.01667 kg)
Zollpfund		
(Germany)	all products	30 Zollloth (0.500 kg)
Zollzentner		
(Germany)	commercial goods	100 Zollpfund (50 kg)

NOTES

Chapter 1

1. For extensive discussions of these various metric system issues, see Chapter 4 ff, Ronald Edward Zupko, "The Origin and Development of the Metric System in France: An Historical Outline," in Metric System Guide Bulletin (XI, 1), 1974, pp. 25-31, and "Worldwide Dissemination of the Metric System During the Nineteenth and Twentieth Centuries," in Ibid, (IX, 2), 1974, pp. 14-25. Throughout the text the designation "metric" appears, even though the name of the system was changed officially more than two decades ago. In order to recognize the universal employment of metric weights and measures, delegates to the 11th General Conference of Weights and Measures in 1960 agreed to change its name to Système International (SI). But since the present work deals with metrological issues beginning approximately in the seventeenth century, it would be misleading to use the modernized name for any developments before 1960. To maintain uniformity throughout, the name "metric" occurs except in those few cases where there is no possibility of confusion.

2. See Ronald Edward Zupko, A Dictionary of English Weights and Measures from Anglo-Saxon Times to the Nineteenth Century (Madison, 1968); A Dictionary of Weights and Measures for the British Isles: The Middle Ages to the Twentieth Century (Philadelphia, 1985); British Weights and Measures: A History from Antiquity to the Seventeenth Century (Madison, 1977); French Weights and Measures Before the Revolution: A Dictionary of Provincial and Local Units (Bloomington, 1978); Italian

Weights and Measures: The Later Middle Ages to the Nineteenth Century (Philadelphia, 1981); "Notes on Medieval English Weights and Measures in Francesco Balducci Pegolotti's 'La Pratica Della Mercatura'," in Economy, Society, and Government in Medieval Italy, 1969, pp. 153-60; "The Weights and Measures of Scotland Before the Union," in Scottish Historical Review, October 1977, (LVI, 2: No. 162), pp. 119-145; and "English Weights and Measures: The Historical Evolution from Roman to Metric Standards," in Technikatorteneti Szemle: Proceedings of the Hungarian Academy of Sciences, Vol. X (1978), pp. 211-221. All of these books and articles contain extensive bibliographies.

In addition, the following works are valuable for their coverage of various aspects of metrological variation throughout the European continent: Wilfrid Airy, "On the Origin of the British Measures of Capacity, Weight and Length," in Minutes of Proceedings of the Institution of Civil Engineers, 175 (1909), pp. 164-176; John Henry Alexander, Universal Dictionary of Weights and Measures (Baltimore, 1850); L. Amati, Weights, Measures and Interest Tables (Milan, 1891); Josef Aubok, Hand-Lexikon über Münzen, Geldwerthe, Tauschmittel, Zeit-, Raum- und Gewichtsmasse der Gegenwart und Vergangenheit aller Länder der Erde (Vienna, 1893); A.E. Berriman, Historical Metrology (London, 1953); L.C. Bleibtren, Handbuch der Münz = Maass = und Gewichtskunde (Stuttgart, 1863); August Blind, Mass-, Münz- und Gewichtswesen (Leipzig, 1906); Walter Block, Masse und Messen (Leipzig, 1913); H.A.

Chapter 1

Bunting, *The Standard English & Foreign Calculator of Money, Weights & Measures* (Manchester, 1906); Frank Wigglesworth Clarke, *Weights, Measures and Money of All Nations* (New York, 1888); Layton Cooke, *Tables Adapted to the Use of Farmers and Graziers* (London, 1819); Carl Cruger, *Contorist: Eine Handels- Münz- Maass- und Gewichtskunde* (Hamburg, 1830); Ezekiel B. Elliott, *Tables of Money, Weights, and Measures of the Principal Commercial Countries in the World* (New York, 1869); Franz Engel, *Tabellen alter Münzen, Masse und Gewichte* (Rinteln, 1965); Matthew D. Finn, *The Commercial Adjuster of Foreign Money, Weights and Measures* (New York, 1843); Roland Goock, *Messen, Wiegen, Zählen: Das Lexicon der Mass- und Währungseinheiten aller Zeiten und Länder mit über 2000 Stickwörtern* (Gütersloh, 1971); William Gutteridge, *A Set of Tables of All the Measures of Capacity Used, Generally and Provincially, within the Dominions...of the British Empire* (London, 1825); Richard Hayes, *The Negociator´s Magazine: or, The Most Authentic Account...of the Monies, Weights, and Measures of the Principal Places of Trade in the Known World* (London, 1777); Johann Peter Heuser, *Ueber bürgerliche Masse und Gewichte* (Elberfeld, 1839); Highland and Agricultural Society of Scotland, *Report on Weights and Measures* (Edinburgh, 1813); Alexander Huntar, *A Treatise of Weights, Mets and Measures of Scotland* (Edinburgh, 1624); Alexander Justice, *A General Treatise of Monies and Exchanges* (London, 1707); Patrick Kelly, *The Universal Cambist and Commercial Instructor* (London, 1821); Richard Klimpert, *Lexicon der Münzen, Mässe, Gewichte: Zählarten und*

Zeitgrössen aller Länder der Erde (Berlin, 1896); Ferdinand Klotz, *Münzen, Maassen und Gewichte* (Würzburg, 1843); Jürgen Elert Kruse, *Allgemeiner und besonders Hamburgischer Contorist* (Hamburg, 1784); Angelo Martini, *Manuale di metrologia* (Turin, 1883); Stephen Naft, *Conversion Equivalents in International Trade* (Philadelphia, 1931); Christian Noback, *Münz-, Maass- und Gewichtsbuch* (Leipzig, 1858); Egon S. von Oelsen, *Währungen, Masse, Gewichte der ganzen Welt* (Vienna, 1933); Joseph Palethorpe, *A Commercial Dictionary of the Names of All the Coins, Weights and Measures in the World* (Derby, 1829); Christopher Knight Sanders, *A Series of Tables in Which the Weights and Measures of France Are Reduced to the English Standard* (London, 1825); C.V.J. Scherer, *Allgemeiner Contorist* (Hamburg, 1834); Friedrich Heinrich Schlössing, *Handbuch der Münz- Mass- und Gewichtskunde* (Stuttgart, 1890); Frederick George Skinner, *Weights and Measures: Their Ancient Origins and Their Development in Great Britain up to AD 1855* (London, 1967); William Tate, *The Modern Cambist: Forming a Manual of Foreign Exchanges...with Tables of Foreign Weights and Measures* (London, 1849); Guérin de Thionville, *Tavole delle monete, pesi e misure dei principali paesi dei globo, e de´ principali popoli dell´ antichità* (Naples, 1848); Vincenzo Tonarini, *Ragguagli dei cambj, pesi, e misure delle più mercantili piazze di Europa* (Bologna, 1780); Antonio Maria Triuizi, *Bilancio dei pesi e misure di tutte le piazze mercantili dell´Europa* (Venice, 1803); and "Weights, Measures, and Money of the Various Countries with Their English Equivalents," in *International Rural*

Scientific and Practical Agriculture, 9 (January, 1918), pp. 9-11.

3. See Ronald Edward Zupko, "Western European Weights and Measures," in Dictionary of the Middle Ages (New York, vol. 12, forthcoming), and other articles on metrology and numismatics throughout the twelve volumes.

4. See Appendix 1 for a complete listing of all of these units and their relationships to standard English units. Also see Ronald Edward Zupko, "Medieval English Weights and Measures: Variation and Standardization," in Studies in Medieval Culture, 4 (1974), pp. 238-243.

5. In the historical development of metrological systems, quantity measures evolved principally as a method for bulkrating industrial, commercial, and agricultural products. Representing the number or count of any item or set of items, this branch of measurement has always been employed whenever the physical attributes of certain products precluded their easy reckoning, classification, or sale by measures of length, capacity, volume, or weight. Examples from manufacturing or industry would be combs, pipes, ropes, brooms, knives, gloves, plates, reeds, tile, bricks, oars, staves, bristles, hoops, boards, nails, pins, and poles. In agriculture many seeds, grains, grain products, vegetables, fruits, cattle, and animal products fell under this category. Merchants depended on quantity measures for buying and selling items such as inexpensive spices and textiles, common medical, culinary, and pharmaceutical supplies, raw materials,

naval stores, fish, animal skins, furs, pelts, lumber, firewood, ores, metals, minerals, heating materials, clothes, paper products, dinnerware, and glassware. Many other occupations and products could be cited.

Further, quantity measures became vitally important after various European governments began to establish metrological standardization programs during the later Middle Ages. These programs aimed at limiting the numbers and varieties of weights and of linear and capacity measures in actual use, and at establishing acceptable national dimensions and applications for them. In other words, governments wanted all local units to conform to state standards. Even though most of these standardization attempts were unsuccessful prior to the creation and dissemination of the metric system throughout Europe during the nineteenth century, they did curtail to a considerable degree the proliferation of new weights and measures to handle the needs of an expanding industrial, mercantile, and agricultural world. Quantity measures were substituted for others increasingly after the Industrial Revolution since groups or sets of acceptable, established units could be combined in varying numbers and proportions for almost any need. There were no physical standards needed for new quantity measures, only new mental conceptions.

6. See Appendix 2 for a thorough listing of the quantity measures of pre-metric Europe.
7. Demi-measures also existed for the baril, barrique, boisseau, brasse,

caque, carat, gros, lieue, litron, livre, minot, muid, once, pièce, pipe, posson, quartaut, quart(e), quarteranche, queue, quintal, roquille, scrupule, setier, somme, and voie. Measures such as the demion and demoiselle also can be included in this category.

8. The most important were the quartal (= 1/4 baril), quartaut (= 1/4 muid or queue), quarte (= 1/4 pot, velte, or setier), quartée (= 1/4 of any land unit), quartel (= 1/4 of some larger land unit), quartelade, quartelée (= 1/4 arpent or mine), quartelet (= 1/4 quartc), quarteranche (= 1/4 bichetée), quarterée, quarternel, quarteron (= 1/4 livre poids de marc, cent, or centaine), quartier (= 1/4 setier), quartière (= 1/4 émine), and quartonnier (= 1/4 boisseau).

9. Common among the French series were tiercel (= 1/3 arpent), tiercelée (= 1/3 setier), tierceron, tiercière, tierçon (= 1/3 muid), and tierçuel. England had the tierce (= 1/3 pipe) and the tertian (= 1/3 tun of 252 gallons, synonymous with the puncheon, and double the tierce of 42 gallons).

10. France formed diminutives with the suffixes -lot (barillot), -sel (barisel), -let (bariselet), -el (pintel), -elle (coupelle), ette (pintelette), and -on (peson). In the British Isles -et was the standard suffix as in balet, while in Italy -etta (balletta), -ciello (balonciello), -ino (quanottino), and -otto (granotto) were used.

11. The German numerical prefixes were employed commonly to express parts or percentages of the following major units: linear (Elle, Fuss, Klafter, Linie, Meile, Ruthe, and Zoll); area (Morgen); volume

(Klafter); liquid capacity (Anker, Eimer, Fass, Fuder, Kanne, Mass, Ohm, Oxhoft, and Tonne); and dry capacity (Himt, Malter, Metze, Scheffel, Tonne, Viertel, and Wispel).

12. In Hamburg the Palme was considered equal to 1/3 Fuss or 4 Zoll or 9.55 cm. The submultiples were similar everywhere in western and eastern Europe.

13. In Hannover and Brunswick the Spann was considered equal to 10 Lachterzoll or 2.40 dm.

Chapter 2

1. These problems were commonplace throughout all European states before the modern era.

2. For fuller discussions of these decrees and legislative enactments see Ronald Edward Zupko, British Weights & Measures: A History from Antiquity to the Seventeenth Century (Madison, 1977), pp. 16-93. The principal English documents in question were the Magna Carta (1215), Assize of Bread & Ale (1266), Composition of Yards & Perches (1266-1303), Decree of 1296, Tractatus (1303), Ordinance for Measuring Land (1305), Act of 1324, Ordinance for Measures (1325), Statute of Purveyors (1351), Act of 1351, Statute of the Staple (1353), Acts of 1353, 1357, Statute of Westminster I (1357), and Acts of 1360, 1389, 1391, 1414, 1421, 1423 (I and II), 1430, 1439, 1449, 1482, 1483, 1491, 1495, 1496, 1527, 1531, 1536, 1541, and 1570. In Scotland of major concern were King David's Assize and the Acts of 1435, 1457, 1487,

1503, 1555, 1563, 1573, and 1587.

Among the most important documentary sources for this legislative record, and for that of succeeding ages, are George Adam, "Weights and Measures Legislation," in The Decimal Educator, 14 (1932), pp. 40-43; Acts of the Parliaments of Scotland, 1424-1727 (London, 1908); Acts of the Parliaments of Scotland, 1124-1707, with Supplement, ed. T. Thomson and C. Innes (12 vols., Edinburgh, 1875); Board of Trade, Parliamentary Papers, Great Britain: First Report of the Warden of the Standards on the Proceedings and Business of the Standard Weights and Measures Department of the Board of Trade for 1866-1867 (and reports dated 1868, 1869, 1870, 1871, 1873, 1875, 1876, 1877, 1878, 1879, 1902, 1914, 1915, 1921, 1922, 1925, 1928, 1929, 1930, 1931, 1933, 1934, 1935, 1936, 1937, 1938, 1939, and 1951) (London, 1867 ff); William Eric Bousfield, The Weights and Measures Acts, 1878 to 1904 (London, 1907); Howard Cunliffe, Weights and Measures Act, 1904 (Smethwick, 1913); John Devonald Fletcher, The Weights and Measures Acts 1878 to 1904 (London, 1908); Ireland, Index to the Statutes at Present in Force in, or Affecting, Ireland from the Year 1310 to 1835 (Dublin, 1836); Laws and Acts of Parliament: James VII to Anne (1685-1707) (Edinburgh, 1731); Laws and Acts of Parliament Made by James I, II, III, IV, V, Queen Mary, James VI, Charles I and II (1424-1681), ed. T. Murray (Edinburgh, 1682-1685); John A. O´Keefe, The Law of Weights and Measures (London, 1966) and The Law of Weights and Measures: Supplement (London, 1967); George A. Owen, The Law Relating to Weights and Measures (London,

1947); Scots Statutes Revised, 1424-1900 (10 vols., Edinburgh, 1899-1907); Smithsonian Institution: National Museum of History and Technology, Washington, D.C., British Acts Relating to Weights and Measures: 1824-1871 (Washington, D.C., 1892); John Whitehurst, An Attempt toward Obtaining Invariable Measures of Length, Capacity, and Weight (London, 1787); George Crispe Whiteley, The Law Relating to Weights, Measures, and Weighing Machines (London, 1879); and Charles Moore Watson, British Weights and Measures as Described in the Laws of England from Anglo-Saxon Times (London, 1910).

3. For photographic plates of the pre-Stuart national standards, see Zupko, British Weights & Measures, pp. 76, 79, 80, 86, 89, 90-91.
4. See Zupko, British Weights & Measures, pp. 34-70, 81-86.
5. For extensive discussions of the full range of the technological innovations and inventions of the seventeenth and early eighteenth centuries, and for the detailed inner workings of the scientific societies and astronomical observatories, see Hugh Kearney, Science and Change, 1500-1700 (New York, 1971), pp. 7-12, 19-25, 44-48, 65-66, 141-158, 171-185, 216-217; A. Rupert Hall, The Scientific Revolution: 1500-1800 (London, 1962), pp. 186-204, 234-243; Herbert Butterfield, The Origins of Modern Science: 1300-1800 (New York, 1951), pp. 58-70; Harold G. Bowen and Charles F. Kettering, A Short History of Technology (West Orange, New Jersey, 1954), pp. 49-52; Maurice Daumas, Scientific Instruments of the Seventeenth and Eighteenth Centuries (New York, 1972), pp. 121-135; Marian Card Donnelly, A Short History of

Observatories (Eugene, Oregon, 1974), pp. 3-28; F.A. Towle, "The Royal Society," in London and the Advancement of Science (London, 1931), pp. 40-47; Frank Heath, "Government and Scientific Research," in Ibid., pp. 188-230; Frank Dyson, "The Royal Observatory, Greenwich," in Ibid., pp. 231-234; F.A. Bather, "The Museums of London," in Ibid., pp. 271-300; Robert S. Whipple, "A Brief History of the London Makers of Scientific Instruments," in Ibid., pp. 301-311; C.A. Alexander, "Outline of the Origin and History of the Royal Society of London," in Annual Report of the Board of Regents of the Smithsonian Institution...for the Year 1863 (Washington, D.C., 1864), pp. 137-152; A. Rupert Hall, From Galileo to Newton: 1630-1720 (New York, 1963), pp. 134-152; A. Wolf, A History of Science, Technology, and Philosophy in the 16th & 17th Centuries (London, 1950), pp. 55-70, 165-179; Richard Foster Jones, Ancients and Moderns: A Study of the Rise of the Scientific Movement in Seventeenth-Century England (St. Louis, 1961), pp. 178-226; A. Rupert Hall and Marie Boas-Hall, "Anglo-French Scientific Communication in the Mid-Seventeenth Century," in XIIe Congrès International d'Histoire des Sciences (Paris, 1971), pp. 65-69; A.R.J.P. Ubbelohde, "The Beginning of the Change from Craft Mystery to Science as a Basis for Technology," in A History of Technology, ed. Charles Singer, et al., IV, pp. 667-677; and J.D. Bernal, Science in History (New York, 1965), pp. 251-357.

Other works important in their entirety are J. Bertrans, L'Académie des Sciences et les académiciens de 1666 à 1793 (Paris, 1869); Thomas

Birch, The History of the Royal Society of London (4 vols., London, 1756-1757); Roger Hahn, The Anatomy of a Scientific Institution: The Paris Academy of Sciences, 1666-1803 (Berkeley, 1971); Histoire de l'Académie Royale des Sciences, avec les mémoires de mathématique et de physique (92 vols., Paris, 1702-1797); Margery Purver, The Royal Society: Concept and Creation (Cambridge, Mass., 1967); Thomas Sprat, History of the Royal Society (London, 1667); and C.R. Weld, A History of the Royal Society (2 vols., London, 1848).

6. "An Account of a Comparison Lately Made by Some Gentlemen of the Royal Society, of the Standard of a Yard, and the Several Weights Lately Made for Their Use, etc.," in Philosophical Transactions, 42 (1742-43), p. 554. In The York Mercers and Merchant Adventurers, p. 297, it was reported that the city of York received in 1679 a pair of scales and "small exchequer weights from a pound to a dram" for a total price of 9 shillings.

7. This standard and the linear measures from the Exchequer, the Royal Society, the Guildhall, the Tower of London, and several other locations were compared by F. Baily, "Report on the New Standard Scale of This Society," in Memoirs of the Royal Astronomical Society, 9 (1836), pp. 40-41. The results showed that none of them was identical in length, varying by +0.05 to -0.02 from the Exchequer standard.

8. The four standards are the ale quart, avoirdupois flat 8 pound circular weight, avoirdupois 7 pound woolweight, and grain quart.

9. The "Account of a Comparison Lately Made by Some Gentlemen of the Royal

Society," p. 555, also reported that in the Tower of London was stored a pile of hollow troy weights from 256 ounces to 1/16 ounce. All of them down to 8 ounces were marked with a crown over AR and inscribed PRIMO MAII, A^{O} DNI. 1707. A^{O} REGNI VI°. The 4 and 2 ounce weights were marked only with the crown over AR. The smaller weights contained the Exchequer seal and the rose and crown.

10. The statutes were those of 1618, 1641, 1644, 1649, 1655, 1660, 1661, 1662, 1663, 1685, 1688, 1689, 1692, 1693 (I and II), 1694, 1695, 1696, 1697, 1698, 1699 (I and II), 1700, 1701, 1706, 1708, 1710, 1711, 1713, 1729, and 1758.

11. In the "Report from the Committee Appointed to Inquire into the Original Standards of Weights and Measures in This Kingdom, and to Consider the Laws Relating Thereto," in Report from Committees of the House of Commons, 2 (1737-65), p. 427, there is a list of fees that were customarily charged for assaying the various denominations of weights.

12. Found in The Acts of the Parliaments of Scotland, Vol. III, pp. 437-438, the wording of this enactment, typical of others that created weights and measures commissions, was as follows (the paleographical expansions are mine):

> Mr Dauid makgill of nysbite his hienes aduocate
> Mr Dauid carnegie of culluthie Robert fairlie of
> braid Sr arnold naper of edinbellie knicht generall
> of his hienes cunzehous Johnne arnot commissioner

of edinburgh Williame flemyng commissioner of perth
Robert forrester provest and commissioner of striuiling
and hew campbell provest and commissioner of Irwing
...to convene...within the burgh of edinburgh....
And efter sicht and consideratioun of the lawes and
actis of Parliament maid anentis mettis mesuris and
wechtis in tyme bygane and grounds quhairon that
haif proceidit have and regaird to equitie and in-
differencie To sett mak and establishe ane mett mesour
and wecht ilk salbe commoun and vniuersall amangis
all our soverane lordis lieges.

Chapter 3

1. "Report from the Committee Appointed to Inquire into the Original Standards of Weights and Measures in This Kingdom, and to Consider the Laws Relating Thereto," in Report from Committees of the House of Commons, 2 (1737-1765), pp. 428-429.
2. F. Baily, "Report of the New Standard Scale of This Society," in Memoirs of the Royal Astronomical Society, 9 (1836), p. 35.
3. The standards tested by this Committee totaled 86 in number and were manufactured between 1582 and 1737. They included avoirdupois bell-shaped, flat-shaped, troy, and unspecified weights, linear measures of the yard and ell, and capacity measures consisting of the bushel, half-bushel, peck, half-peck, half-sack, gallon, quart, pint,

and gill.

4. Specifically, the hollow pile consists of 1/2 (solid), 1/2, 1, 2, 3, 6, and 12 ounces; the flat pile of 1 (octagonal), 1, 2, 3, 6, and 12 grains, and 1, 2, 3, and 5 pennyweights.

5. His presentation of points one and two is not dealt with at great length. He only scans the variations for the acre, bushel, mile, pound, gallon, coomb, seam, hobed, faggot, gad, burden, fother, cade, last, dicker, way, clove, boll, batement, windle, hoop, and several other units. His major conclusions concerning the second are that the old standards were defective, poorly constructed or damaged, and possessed neither uniformity nor proportion to one another. He also emphasized that parliamentary and local legislation had done little to prevent the omnipresent threat of corruption and decay from engulfing the system. Perhaps his cautionary stance was necessitated by his fear of alienating local and national government authorities. Politics played as important a role in metrology as in other endeavors.

6. William Emerson (1701-82).

7. John Theophilus Desaguliers (1685-1744).

8. John Whitehurst (1713-88).

9. Eliot, Letters, p. 7.

10. Ibid., p. 8.

11. Ibid., p. 10.

12. Young was a scholar of impressive range: a physicist, mathematician, linguist, and Egyptologist. He was elected a Fellow of the Royal

Society at the age of twenty-one.

13. Kater was another remarkable person. A career army officer and physicist, he assisted in the "great trigonometrical survey" of India, and pioneered in the development of the convertible pendulum as an alternative to the approximation of the simple pendulum for the measurement of the seconds pendulum. His convertible pendulum, and the invariable pendulum introduced by him in 1819, were the basis of English pendulum and other horological work.

14. See "Second Report of the Commissioners Appointed by His Majesty to Consider the Subject of Weights and Measures," in Reports from Commissioners, 7 (1820), pp. 1-40.

15. Among the works that treat some of the overall manifestations of these scientific developments, see "Weights and Measures," in The American Cyclopaedia (New York, 1876), vol. 16, pp. 537-544; W.S.B. Woolhouse, Measures, Weights & Moneys of All Nations (London, 1890), pp. 2-5; Henry Lyons, The Royal Society, 1660-1940 (New York, 1968), pp. 173-175, 204-205, 220-227; Baily, "Report on the New Standard," vol. 9, pp. 37-45; S.C. Walker, "Report on the Weights and Measures of Great Britain," in Report of the Managers of the Franklin Institute (Philadelphia, 1834), pp. 35-43; William Harkness, "The Progress of Science as Exemplified in the Art of Weighing and Measuring," in Smithsonian Miscellaneous Collections (Washington, D.C., 1890), pp. 601-605, 609-610; John Henry Alexander, An Inquiry into the English System of Weights and Measures (Oxford, 1857), pp. 28-37, 85-89; George

Biddell Airy, "Figure of the Earth," in Encyclopaedia Metropolitana (London, 1845), vol. 5, pp. 165-240; Edward Sabine, An Account of Experiments to Determine the Figure of the Earth by Means of the Pendulum Vibrating Seconds in Different Latitudes (London, 1825), pp. 366-372; J.D. Bernal, Science in History (New York, 1965), pp. 358-409; and Shepard B. Clough and Richard T. Rapp, European Economic History: The Economic Development of Western Civilization (New York, 1975), pp. 233-236, 291-294.

Chapter 4

1. For the enormous number of unit variations used throughout France and its possessions, see Ronald Edward Zupko, French Weights and Measures Before the Revolution: A Dictionary of Provincial and Local Units (Bloomington, Indiana, 1978), pp. 1-183; "Lettre du Secrétaire de l´Académie des Sciences, 12 juin 1792," in Oeuvres complètes de Condorcet (Paris, 1804), vol. 18, pp. 180-194; Collection complète des lois, décrets, ordonnances, réglemens, avis du Conseil d´état, publiée sur les éditions officielles du Louvre...de 1788 à 1830 inclusivement, ed. J.B. Duvergier (Paris, 1834), vols. 2, 9, 11, 17, 21, and 25; Adrien Fauve, Les Origines du système métrique (Paris, 1931), pp. 1-16; M.A. Grivel, Les anciennes Mesures de France, de Lorraine & de Remiremont, (Remiremont, 1914), pp. 1 ff; and Armand Machabey, Aspects de la métrologie au XVII$^{\underline{e}}$ siècle (Paris, 1954), pp. 5-18.

Other works of significance for the study of French pre-metric measuring units include: C. Afanassiev, Tableau des mesures pour les grains qui étaient en usage en France au XVIII siècle (Odessa, 1891); T. Altès, Traité comparatif des monnaies, poids et mesures, changes, banques et fonds publics entre la France, l'Espagne et l'Angleterre (Marseille, 1832); Jean Baptiste Anville, Traité des mesures itinéraires anciennes et modernes (Paris, 1769); F. Bailly, "Notice sur les anciennes mesures de Bourgogne," in Société d'histoire, d'archéologie et de littérature de l'arrondissement de Beaune, (1902): 173-223, (1903): 156-210, (1904): 177-265, (1905): 223-306; M. Barreme, Le Livre des comptes-faits ou tarif général des monnoyes (Paris, 1755); Anatole J.B.A. Barthelémy, Nouveau Manuel complet de numismatique du moyen age et moderne (Paris, 1852); N. Binet, Tarifs pour les réductions et évaluations des aunes et mesures étrangères en aunes de Paris (Paris, 1698); Charles Boucaud, Les Pichets d'étain; mesures à vin de l'ancienne France (Paris, 1958); Albert Bourgaux, Dictionnaire international des mesures, poids, monnaies, etc. (Brussels, 1927); N.H. Brelet, Le Traducteur des anciennes mesures en nouvelles (Aurillac, 1840); Mathurin Brisson, Réduction des mesures et poids anciens en mesures et poids nouveaux (Paris, 1798); F. Chailan, Tables pratiques des capacités des segmens des tonneaux destinés au transport des liquides sur les principales places de commerce du monde (Marseille, 1834); Frédéric A. Crichton, Traité des poids, mesures et monnaie anglais avec équivalents métriques, augmenté de nombreuses notes explicatives (Paris, 1900); P. Debures, Tableau complet des poids *et mesures anciennement en usage à Marseille et à Paris* (Marseille, 1802); Maurice Denis-Papin and Jacques Vallot, *Métrologie générale (grandeurs et unités)* (Paris, 1946); Horace Doursther, *Dictionnaire universel des poids et mesures anciens et modernes contenant des tables des monnaies de tous les pays* (Anvers, 1840); S. Durant and Alexandre Bastide, *Tables de comparaison entre les anciens poids et mesures de toutes les communes du département du Gard, et les poids et mesures métriques* (Nismes, 1816); Louis Gaillardie, *Poids anciens des villes de France* (Paris, 1898); A. Grué, *Dictionnaire usuel des poids et mesures, ou guide des acheteurs et des vendeurs* (Paris, 1840); E. Hocquart, *Le Livre des poids et mesures* (Paris, 1848); F. Hutinet, *Concordance des anciens poids et des anciennes mesures...avec les nouveaux poids et les nouvelles mesures* (Troyes, 1840); Robert Latouche, *Les Mesures de capacité en Dauphiné du XIVe siècle à la révolution française* (N.p., 1931); F. Lauradoux, *Comptes faits, ou tableaux comparatifs des anciens poids et mesures qui étaient usités dans le département du Rhône avant le système métrique* (Lyon, 1812); Louis Maurice, *Poids et mesures avec l'indication des monnaies de tous les pays* (Paris, 1850); A. Mauricet, *Des anciennes Mesures de capacité et de superficie dans les départements du Morbihan, du Finistère et des Côtes-du-Nord* (Vannes, 1893); L. Passot, *Tables comparées des anciennes et nouvelles mesures généralement usitées en France* (Paris, 1840); A. Peigné, *Conversion des mesures, monnaies et poids de tous les pays étrangers en mesures, monnaies et poids de la France* (Paris, 1867); Emile Peraud, *Barème ou* comptes faits: Tableaux comparatifs des mesures entre elles, métriques, fédérales et pieds-de-roi (Geneva, 1858); Leon Roche, Poids et mesures (Orange, 1837); Jean Baptiste Romé de L'Isle, Métrologie, ou tables pour servir à l'intelligence des poids et mesures des anciens (Paris, 1789); Jacques Frédéric Saigey, Traité de métrologie ancienne et moderne, suivi d'un précis de chronologie et des signes numériques (Paris, 1834); Edouard de Simencourt, Tableaux des monnoies de change et des monnoies réelles, des poids et mesures, des cours des changes et des usages commerciaux des principales villes de l'Europe (Paris, 1817); Michel C. Soutzo, Nouvelles Recherches sur les origines & les rapport de quelques poids antiques (Paris, 1895); and E. Thoison, "Recherches sur les anciennes mesures en usage dans le Gâtinais Seine-et-Marnais et sur leur valeur en mesures métriques," in Bulletin historique et philologique du comité des travaux historiques et scientifiques, (1903), pp. 328-406.

2. A complete transcription of the principal acts, decrees, and orders dealing with pre-metric French weights and measures can be found in Recueil général des anciennes lois françaises, depuis l'an 420 jusqu' à la Révolution de 1789, ed. M. Jourdan, et al. (Paris, 1830), vols. 1, 3, 9, 12, 13, 20, 22, 24, 25, 26, and 27. Besides numerous calls for uniformity of weights and measures and for prohibitions against frauds, these laws yield a great deal of information on many pre-metric units.

3. For a listing and description of the pre-metric national, provincial, and local standards, together with many photographic plates, see

Conservatoire National des Arts et Métiers, _Catalogue du Musée, Section K: Poids et Mesures, Métrologie_ (Paris, 1941), pp. 31-38, 63-73, and 103-113.

4. Speaking of the chaotic state of French weights and measures during the eighteenth century, F. Gattey, _Eléments du nouveau système métrique, suivis des tables de rapports des anciennes mesures agraires avec les nouvelles_, (Paris, 1801), pp. 1-2, wrote that the weights and measures of the ancient regime differed by province, by city, and by village, and that they had different names, different values, different functions, and different applications depending on social, economic, and political usage. He saw total confusion and fraud as running rampant everywhere in the kingdom.

5. See Henri Moreau, _Le Système métrique_ (Paris, 1975), pp. 11-17 and Hans-Joachim von Alberti, _Mass und Gewicht: Geschichtliche und tabellarische Darstellungen von den Anfängen bis zur Gegenwart_ (Berlin, 1957), pp. 88-90. The Grand Châtelet was a fortress that, along with the Petit Châtelet, guarded two bridges providing access to the city.

6. Most of the pre-metric French linear standards were iron bars having their ends turned up at right angles to form talons. The verification of local and regional end measures was accomplished by fitting them between the talons. This type of standard was never employed in England.

7. The actual standard was not very precise based upon the definition. Abbot Picard compared it with the standard fathom and found it to be 3

feet, 7 inches, 10 4/5 lines. Charles Dufay in 1736 got 3 feet, 7 inches, 10 5/6 lines; so did a team of academicians in 1745. Thus, its actual length was about 2.6 lines (6 mm) longer than the statutory definition.

8. A government report of June 17, 1799, attributes its construction to the fourteenth century. Later authorities place it in the last third of the fifteenth century.

9. Excellent analyses of the early history of pendulums and pendulum experiments may be found in Olinthus Gregory, A Treatise of Mechanics, Theoretical, Practical, and Descriptive (London, 1826), vol. 2, pp. 286-293; Observations de la Société Royale d'Agriculture sur l'uniformité des poids et des mesures (Paris, 1790), pp. 10-16, 109; Pierre Mesnage, "The Building of Clocks," in A History of Technology and Invention, ed. Maurice Daumas and trans. Eileen B. Hennessy (1969), vol. 2, pp. 283-305; H. Alan Lloyd, "Mechanical Timekeepers," in A History of Technology, ed. Charles Singer, et al. (Oxford, 1955-1958), vol. 3, pp. 662-675; Thomas Young, "On Drawing, Writing, and Measuring," in A Course of Lectures on Natural Philosophy and the Mechanical Arts (London, 1845), vol. 1, pp. 71-87; Victor F. Lenzen and Robert P. Multhauft, "Development of Gravity Pendulums in the 19th Century," in Contributions from the Museum of History and Technology (Washington, D.C., 1966), Paper 44, pp. 302-347; Patrick Kelly, Metrology: or, An Exposition of Weights and Measures, Chiefly Those of Great Britain and France (London, 1816), p. 8; and Henry Kater and

Dionysius Lardner, Treatise on Mechanics (Philadelphia, 1838), pp. 123-135, 259-287.

For the geodetic operations on the meridian arc, see Giorgio Abetti, The History of Astronomy (New York, 1952), p. 147; Miguel Merino, "Figure of the Earth," in Smithsonian Institution: Astronomy and Mathematics, 1853-1874 (Washington, D.C., 1874), pp. 306-330; Oliver Justin Lee, Measuring Our Universe: From the Inner Atom to Outer Space (New York, 1950), pp. 16-19; Georges Perrier, Petite Histoire de la géodésie: Comment l'homme a mesuré et pesé la terre (Paris, 1939), pp. 16-50; Denison Olmsted, "The Figure of the Earth," in Letters on Astronomy (Boston, 1840), pp. 69-80; W.A. Browne, The Merchants' Handbook of Money, Weights and Measures with Their British Equivalents (London, 1899), pp. 292-297; Edward Sabine, An Account of Experiments to Determine the Figure of the Earth (London, 1825), pp. 364-372; Elias Loomis, A Treatise on Astronomy (New York, 1865), pp. 9-51; François Arago, Astronomie populaire (Paris, 1856), vol. 3, pp. 310-341; Pierre Simon Marquis de Laplace, Exposition du Système du monde (Paris, 1813), pp. 60-82; R.A. Skelton, "Cartography," in A History of Technology, ed. Charles Singer, et al., vol. 4, pp. 596-628; E.G.R. Taylor, "Cartography, Survey, and Navigation 1400-1750," in Ibid, vol. 3, pp. 537-544; Jean Baptiste Joseph Delambre, Grandeur et figure de la terre (Paris, 1912), pp. 199-255; A. Caswell, "Lectures on Astronomy: The Figure and Magnitude of the Earth," in Annual Report of the Board of Regents of the Smithsonian Institution...for the Year 1858 (Washington,

D.C., 1859), pp. 85-104; George Fisher, "On the Figure of the Earth, as Deduced from the Measurements of Arcs of the Meridian and Observations on Pendulums," in The Quarterly Journal of Literature, Science and the Arts (London, 1819), vol. 7, pp. 299-312; Arthur D. Butterfield, A History of the Determination of the Figure of the Earth from Arc Measurements (Worcester, Mass., 1906), pp. 7-88, 151-168; and George Biddell Airy, "Figure of the Earth," in Encyclopaedia Metropolitana: or Universal Dictionary of Knowledge, ed. Edward Smedley, et al. (London, 1845), vol. 5, pp. 165-240.

Other valuable sources in their entirety for both of these measurement standards are Pierre Bouguer, La Figure de la terre, déterminée par les observations de Messieurs Bouguer et de La Condamine, envoyés par ordre du roy au Perou, pour observir aux environs de l´équateur (Paris, 1749); P.L. Moreau de Maupertuis, La Figure de la terre déterminée par les observations de Messieurs de Maupertuis, Clairaut, Camus, Le Monnier, l´Abbé Outhier et Celcius, faites par ordre du roy au circle polaire (Paris, 1738); A.M. Legendre, Méthode pour déterminer la longueur exacte du quart du méridien, d´après les observations faites pour la mesure de l´arc compris entre Dunkerque et Barcelone (Paris, 1799); Charles Marie de La Condamine, Journal du voyage fait par ordre du roi a l´équateur, servant d´introduction à la mesure des trois premiers degrés du méridien (Paris, 1751); Isaac Todhunter, A History of the Mathematical Theories of Attraction and Figure of the Earth, from the Time of Newton to That of Laplace (London, 1873); John Henry Pratt, A Treatise on Attractions, Laplace's Functions, and the Figure of the Earth (London, 1871); James Howard Gore, Elements of Geodesy (New York, 1900); Jean Picard, Degré du Méridien entre Paris et Amiens (Paris, 1740); and Pierre Simon de Laplace, Méchanique céleste, trans. Nathaniel Bowditch (Boston, 1829-1839), 4 vols.

10. On the origin and development of decimals, see Auguste Benoit, Anciennes Mesures d'Eure-et-Loir (Chartres, 1843), pp. 67-70; John Bowring, The Decimal System in Numbers, Coins, and Accounts (London, 1854), chap. 5 ff.; Philipp Lenard, "Simon Stevin," in Great Men of Science: A History of Scientific Progress (New York, 1933), pp. 20-24; Joseph Fayet, La Révolution française et la science, 1789-1795 (Paris, 1960), pp. 450 ff.; Maurice Danloux Dumesnils, Etude critique du système métrique (Paris, 1962), pp. 3 ff.; and Cyril S. Smith and R.J. Forbes, "Metallurgy and Assaying," in A History of Technology, ed. Charles Singer, et al., vol. 3, p. 61.

11. Stevin's monograph was translated into English by R. Norton and published in London in 1608 under the title Disme: The Art of Tenths, or Decimall Arithmetike. Also see George Sarton, "Decimal Systems Early and Late," in Osiris (vol. 9, 1950), pp. 581-601 and "The First Explanation of Decimal Fractions and Measures (1585), Together with a History of the Decimal Idea and a Facsimile (No. XVII) of Stevin's Disme," in Isis (vol. 33, 1935), pp. 153-244.

12. Gabriel Mouton, Observationes diametrorum solis et lunae...huic adjecta

est brevis dissertatio...nova mensurarum geometricarum idea (Lyon, 1670).

13. Some of Mouton´s written observations concerning his early conclusions are recorded in James Howard Gore, "The Decimal System of Measures of the Seventeenth Century," in American Journal of Science, vol. 41 (1891), pp. 242-246. Later statements of praise for Mouton´s scientific and metrological expertise made by Picard between 1672-1674, Cassini in 1757, and Condamine in 1776 can be found here also. See also Robert A. Hopkins, The International (SI) Metric System and How It Works (Tarzana, Calif., 1974), pp. 10-11.

14. During the long and often arduous years of his geodetical experiments, Giacomo Cassini also improved the meridian sundial at the Paris Observatory.

15. Approximately 20 years later Cesare F. Cassini, Giacomo´s son, and other scientists, verified the meridian of France by means of a triangulation that was also employed for the tracing of a complete topographic map of France, commonly called thereafter "Cassini´s map."

16. For analyses of the early years of the history of metric trials and errors, see: L.P. Abeille and M. Tillet, Observations de la Société Roy. d´Agric. sur l´uniformité des poids et des mesures (Paris, 1790); L. Auger, L´Indicateur des poids et mesures métriques (Paris, 1842); Ferdinand Berthoud, Traité des montres à longitudes, contenant la construction, la description & tous les détails de main-d´oeuvre de ces machines; leur dimensions, la manière de les éprouver, etc. (Paris,

1792; R. Bonne, Principes sur les mesures en longueur et en capacité, sur les poids et les monnoies, dépendant du mouvement des astres principaux et de la grandeur de la terre (Paris, 1790); A.J.F. Bossut, Adresse à l'Assemblée Nationale sur les préalables à la pratique d'une seule mesure en ce qui est relatif aux terrains (Paris, 1791); H. Bovy, Traité complet des poids et mesures (Paris, 1839); A.F. Broc, Nouveau Code des poids et mesures (Paris, 1834); Bureau des Longitudes, Paris, Système métrique: Rapport de la commission composée de MM. Mathieu, Laugier et Faye, rapporteur (Paris, n.d.); J.B. Castille, Traité sur le nouveau système des poids et mesures, etc. (Paris, 1801); Léon Chauvin, Histoire du mètre, d'après les travaux et rapports de Delambre, Méchain, Van Swinden, etc. (Limoges, 1901); J. Henri Dugué, Instruction sur les nouvelles mesures et sur la décimal (Le Mans, 1802); J.D. Collene, De la Numération décimale et du système métrique (Paris, 1839) and Le Système octaval, ou la numération et les poids et mesures reformés (Paris, 1840); Christophe Dausse and P. Tremblay, Système nouveaux des poids et mesures (Grenoble, 1823); J.B.J. Delambre, Base du système métrique décimal (3 vols., Paris, 1807-1810); Département de l'Interieur, Paris, Circulaires, instructions et autres actes émanés du Ministère de l'Intérieur...à 1821 (2 vols., Paris, 1821-22); A. Ernaux, Manuel complet du système métrique, appliqué aux nouvelles mesures (Paris, 1839); M. Fort (or St. Pons), Tables de comparaison entre les anciens poids et mesures du département de l'Hérault et les nouveaux poids et mesures (Montpellier, 1799); J.A. Gaillard, Tables de *réduction des anciennes mesures en nouvelles...avec divers documens relatifs au système métrique décrété les 1er août 1793 et 18 germinal an III, etc.* (Le Havre, 1799); F. Gattey, *Tables de réduction des anciens poids en nouveaux et réciproquement à l'usage des pharmacies* (Paris, 1801); J.L. Gautier, *Raisonnement abrégé du système métrique suivi de tables complètes de conversion des mesures agraires et du mesurage pratique de surface et de cube* (Paris, 1841); L.D. Guyot, *Tables et constructions pour opérer facilement dans tous les départements de la France et dans les pays étrangers la conversion des anciennes mesures en celles du système métrique* (Angers, 1801); William Hallock and Herbert T. Wade, *Outlines of the Evolution of Weights and Measures and the Metric System* (New York, 1906); René Just Abbé Haüy, *Instruction sur les mesures déduites de la grandeur de la terre, uniformes pour toute la république, et sur les calculs relatifs à leur division décimale* (Paris, 1795); Henry Hennessy, "Unit of Length, etc.," in *Franklin Institute Journal*, 72 (1861), pp. 346-351; John F.W. Herschel, "The Yard, the Pendulum, and the Metre," in *Familiar Lectures on Scientific Subjects*, pp. 419-451 (New York, 1872); Walter Hough, "The Origin and Development of Metrics," in *American Anthropologist*, New Series, 35 (1933), pp. 443-450; *Instruction sur la manière de rectifier les tables de comparaison entre les anciennes et les nouvelles mesures, etc.* (Paris, 1801); Antoine Laurent Lavoisier, *Oeuvres*, ed. René Fric (2 vols., Paris, 1955); *Le Manuel républicain: Première partie* (Paris, 1799); J.F. Lesparat, *Métrologies* constitutionelles et primitives comparées entre elles et avec la métrologie d´ordonnances (Paris, 1801); P.L. Lionet, Manuel du système métrique (Lille, 1820); C.J.A. Mathieu de Dombasle, Du Système métrique des poids et mesures (Paris, 1837); Ministère du Commerce et de l´Industrie, Paris, Le Système métrique décimal: Sa création en France, son évolution, ses progrès (Paris, 1930); A. Morin, "Notice historique sur le système métrique, sur ses développements et sur sa propagation," in Annales du Conservatoire (Imperial) des Arts et Métiers, 9 (1870), pp. 573-640; Achille Nouhen, Nouveau Manuel comptes-faits ou barème général des poids et mesures mis à la portée de tout le monde (Paris, 1840); J.F.G. Palaiseau, Exposition du système métrique des poids, mesures et monnaies françaises, appliqué au calcul décimal (Paris, 1831) and Métrologie universelle ancienne et moderne ou rapport des poids et mesures des empires, royaumes, duchés et principautés des quatre parties du monde (Bordeaux, 1816); Alfred Perot, The Decimal Metric System: Foundation, International Organization, Future Development (Paris, 1915); Louis Pouchet, Métrologie terrestre ou tables des nouveau poids et mesures (4th ed., Rouen, 1798) and Tableau des nouveaux poids, mesures et monnaies de la République française, etc. (Rouen, 1795); Claude Antoine Prieur (Du Vernois), Instruction sur le calcul décimal appliqué principalement au nouveau système des poids et mesures (Paris, 1795) and Mémoire sur la nécessité et les moyens de rendre uniformes dans le royaume toutes les mesures d´entendue et de pesanteur (Paris, 1790); Louis Puissant,

<u>Réflexions sur les avantages du calcul décimal dans les mesures géométriques</u> (Agen, 1797); A. Réville, <u>Guide pratique des poids et mesures et du système décimal</u> (Paris, 1839) and <u>Nouveau Livre du cubage ou tables métriques, etc.</u> (Le Harve, 1840); Edwin W. Schreiber, "The Early History of the Metric System," in <u>The Metric System of Weights and Measures</u> (The National Council of Teachers of Mathematics: Twentieth Yearbook, New York, 1966); Gustave Tallent, <u>Histoire du système métrique</u> (Paris, 1910); Charles Maurice de Talleyrand-Périgord, <u>Proposition faite à l'Assemblée Nationale, sur les poids et mesures, etc.</u> (Paris, 1790); Sébastien André Tarbé des Sablons, <u>Nouveau Manuel complet des poids et mesures</u> (Paris, 1845); E.A. Tarnier, <u>Livret explicatif des tableaux du système métrique</u> (Paris, 1865); M. Tisseband, <u>Instruction sur le système métrique: Conversion des anciennes mesures de Paris et des départements en mesures nouvelles, etc.</u> (Douai, 1838); and S. Wild, <u>Essai sur une mesure universelle</u> (Lausanne, 1801).

17. An excellent compilation of all the major metric legislation during the crucial period from 1790 to 1821 may be found in <u>Collection complète des lois, décrets, ordonnances, réglemens, avis du conseil d'état, publiée sur les éditions officielles du Louvre</u>, ed. J.B. Duvergier (Paris, 1834), vols. 1, 2, 6, 8, 9, 10, 12, 18, 21, 25, and 27. Good accounts of the various stages in the development of the metric system in France, and its dissemination to other countries, may be found in <u>Encyclopédie moderne: Dictionnaire abregé des sciences, des lettres,</u>

des arts, de l'industrie, de l'agriculture et du commerce (Paris, 1849), vol. 20, pp. 595-605; *Dictionnaire technologique, ou Nouveau Dictionnaire universel des arts et métiers, et de l'économie industrielle et commerciale* (Paris, 1828), vol. 13, pp. 271-299; H. Arthur Klein, *The World of Measurements* (New York, 1974), pp. 105-126; Gerald J. Black, *Canada Goes Metric: With an Introductory History of Measurement* (Toronto, 1974), pp. 85-105; Armand Machabey, "Techniques of Measurement," in *A History of Technology and Invention*, vol. 2, pp. 306-343; Walter B. Scaife, "The Origin of the Metric System," in *Scientific American Supplement* (New York, July-December 1889), vol. 28, pp. 11500-11501; Joseph Fayet, *La Révolution française et la science, 1789-1795* (Paris, 1960), pp. 442-483; and H. Cavalli, *Tables de comparaison des mesures, poids et monnaies anciens et modernes des principales villes commercialles et des plus importantes nations du monde également comparées avec le système métrique moderne* (Marseille, 1869), pp. 9 ff.

18. Jacques Necker, "Compte rendu au Roi de 1778," in G. Bigourdan, *Le Système métrique des poids et mesures* (Paris, 1901), p. 11.

19. John Riggs Miller, *Speeches in the House of Commons upon the Equalization of the Weights and Measures of Great Britain* (London, 1790), p. ii. Miller gave three addresses concerning this reform plan on July 25, 1789, February 6, 1790, and April 13, 1790.

20. On Jefferson's metrological contributions, and on his views concerning reform, see Brooke Hindle, *David Rittenhouse* (Princeton, 1964), pp.

312-317; E. Millicent Sowerby, comp., Catalogue of the Library of Thomas Jefferson (Washington, D.C., 1955), vol. 4, pp. 49-64; John P. Foley, ed., The Jeffersonian Cyclopedia, a Comprehensive Collection of the Views of Thomas Jefferson (New York, 1900), vol. 2, pp. 830-832; Thomas Jefferson, "Plan for Establishing Uniformity in the Coinage, Weights and Measures of the United States: Communicated to the House of Representatives, July 13, 1790," in The Complete Jefferson, ed. Saul K. Padover (New York, 1943), pp. 974-995; and "Standards of Measures, Weights and Coins," in Ibid, pp. 1004-1011.

21. Congress not only would eschew this plan, but they later rejected the imperial system as well.

22. Jefferson did have one success, however, during this period. In 1782 he and Robert Morris issued several reports advocating currency reform in the United States. They selected the dollar as the basic unit of account, with its multiples and submultiples derived decimally. On July 6, 1785, Congress adopted the dollar as the basic unit of coinage, and on August 8, 1786, Congress adopted a complete decimal system of coinage. Great Britain would not decimalize its coinage until almost two centuries later.

23. In a very hostile tone, W. de Fonvielle, La Mesure du mètre; dangers et aventures des savants qui l´ont déterminée (Paris, 1886), remarked that Great Britain had a despotic monarchy, and that this influenced its refusal to consider the metric system. He also accused the English government of being jealous. Aside from his monarchic bias, he may

have been correct.

24. Following Washington's third message to Congress on October 5, 1791, a select Senate Committee on Weights and Measures was appointed, and on April 5, 1792, recommended the adoption of a decimal system of weights and measures, being in essence the second of Jefferson's proposals made in 1790. No legislation resulted, despite the fact that the subject of weights and measures was before the Senate on four separate occasions. Then on January 9, 1795, following a communication from the Foreign Minister of the French Republic, a Select Committee of the House of Representatives was appointed to consider the possible adoption of the metric system, and to reconsider Jefferson's 1790 report. On April 12, 1796, the committee reported in favor of retaining the existing foot and avoirdupois pound systems.

25. For the entire decree, see E. Clémenceau, Le Service des poids et mesures en France à travers les siècles (Paris, 1909), pp. 183 ff. and Collections des décrets de l'Assemblée nationale constituante, ed. M. Arnoult (Dijon, 1792), vol. 2, pp. 281-284.

26. Since the earth is divided into 24 equal time zones, each time zone centers around a meridian or great circle drawn around the earth from the North to the South Poles. These 24 meridians are separated by 15° (or 360° for the entire circumference of the earth). A quadrant is thus 1/4 of the circumference of one of these meridians. In the French experiments the meridian transected a line near Paris.

27. Clémenceau, p. 183, remarked that Méchain was arrested because his

instruments were suspect and he was thought to be a spy.

28. Borda is often given credit for selecting the name "meter," derived from the Greek metron, measure. Since all other measures in this new metrology were based in some way on the meter, it only seemed logical to call the entire creation the "metric system."

29. In Sur l'Uniformité et le système générale des poids et mesures (Paris, 1793), p. 16, Abrogast is quoted as saying that the needs of society were not yet advanced to the point that it was necessary to select the missing metric names; they would come later. Metrological expertise notwithstanding, on this issue history would prove him to be correct. The "methodical" nomenclature was selected for two reasons: it would not compromise the successful establishment of uniform measures, and if the system were to become universal, it could not be assigned to a French or any other contemporary culture or language.

Chapter 5

1. For important discussions of this later phase of metric history, see M. Aliamet and J.A. Montpellier, Guide pratique de mesures et essais industriels (Vol. 1, Paris, 1899); Raymond Allard, Le Système international de mesures (Paris, 1965); G. Allix, Explication d'un nouveau système de tarifs, ou nouvelles methodes pour trouver en mesures métriques, sans aucun calcul, etc. (Paris, 1840); Pamela Anderton and P.H. Bigg, Changing to the Metric System: Conversion Factors, Symbols and Definitions (London, 1969); Anglo-French Ready

Reckoner: Being Tables for the Conversion (Progressively) of French Prices, Weights, and Measures into English and of English into French (Manchester, 1905); E. Anthony, An Enquiry into an Explanation of Decimal Coinage and the Metric System of Weights and Measures (London, 1906); J. Antrade, Le Mouvement: Mesures de l´étendue et mesures de temps (Paris, 1911); Nathan Appleton, Projet d´un système métrique monétaire international pour l´or, l´argent et le cuivre (Paris, 1880); L. Bailly, Application définitive des principes du système métrique décimal aux questions concernant la mesure des angles, la mesure du temps et le calendrier (Paris, 1890); Emile Barbieux, La Législation française des poids et mesures (Paris, 1926); Jean Baret, Tableau des unités de mesure (Paris, 1938); Jean Bernot, Echelles de conversion des unités anglo-saxonnes en unités métriques (Paris, 1949); C. Bopp, Die international Mass-, Gewichte- und Münz- Einigung durch das metric System (Berlin, 1869); J.J. Bourgeois, Nouveau Manuel des poids et mesures (2 vols., Paris, 1879); Jacques Boyer, The Centenary of the Metric System, trans. William H. Seaman (Chicago, 1901); H.W. Chisholm, "The International Metric Commission," in Nature, 7 (1873), pp. 197-198 and "Metric Commission, International, at Paris," in Nature, 13 (1876), pp. 452-454; L. Dalechamps, Manuel des poids et mesures (Paris, 1851); Maurice Danlous-Dumesnils, Le Mètre et les mesures de longueur (Paris, 1967); George W. Emonts, Tables for Computing Equivalent Metric and Non-Metric Weights and Measures (Philadelphia, 1883); Joseph Garnier, Traité des mesures métriques (Paris, 1858); Robert Goltman and Maurice

LeRoy, Le Système métrique (Montreal, 1903); Marvin H. Green, International and Metric Units of Measurement (New York, 1961); J.E. Hilgard, "Report of the Proceedings of the International Standard Commission, and on the Convention Signed at Paris May 20, 1875, for the Establishment of an International Bureau of Weights and Measures," in On the Adoption of the Metric System of Weights and Measures (46th Congress, 1st Session, House of Representatives Report No. 14, pp. 63-66, Washington, D.C., 1879); M. Jacob, Traité complet du nouveau système des poids et mesures (Paris, 1842); Arthur E. Kennelly, Vestiges of Pre-Metric Weights and Measures Persisting in Metric-System Europe, 1926-1927 (New York, 1928); Charles de Laplace, Notice sur le système métrique décimal des poids et mesures (Paris, 1864); J. Michel, Le Centennaire du mètre: Les précurseurs du système métrique et les mesures internationales (Paris, 1898); A.A. Michélson, Valeur du mètre (Paris, 1894); Guilford L. Molesworth, Weights, Measures and Decimal Tables (London, 1932); Maurin Nahuys, Etat de la question de l´uniformité des monnaies, des poids et des mesures (Paris, 1865); "New Metric Standards," in Scientific American Supplement, 39 (January-June, 1895), p. 16075; H.A. Newton, The Metric System of Weights and Measures with Tables (Washington, D.C., 1868); Pierre M. Nichil, Unités de mesure; facteurs, formules de correspondence et de conversion (Paris, 1965); "Régime légal du système métrique dans la monde," in Les récents Progrès du système métrique (1948-1954), pp. 59-71; Arthur T. Shapiro, Le Système métrique (Montreal, 1974); "Third Report of Standards

Commission, February 1, 1870," in Metric System Pamphlets, 1 (1878), No. 2; H. Tresca, "Methodical Statement of the Resolutions Passed by the International Metric Commission During Their Meeting at Paris in 1872," in On the Adoption of the Metric System of Weights and Measures (46th Congress, 1st Session, House of Representatives Report No. 14, pp. 52-55, Washington, D.C., 1879); John Hill Twigg, Summary of Official Reports on the Metric System (London, 1911); United Nations, Department of Economic and Social Affairs, New York, World Weights and Measures Handbook for Statisticians (New York, 1967); United States National Bureau of Standards, Washington, D.C., Brief History and Use of the English and Metric Systems of Measurement (Special Publication No. 304A, Washington, D.C., 1968) and The International Bureau of Weights and Measures, 1875-1975, ed. Chester H. Page and Paul Vigouraux (Special Publication No. 420, Washington, D.C., 1975).

2. See Claude Antoine Prieur (du Vernois), Nouvelle instruction sur les poids et mesures et sur le calcul décimal, adoptée par l'Agence Temporaire des Poids et Mesures (Paris, 1796), pp. 43 ff. One year earlier the Temporary Agency in Avis instructif sur la fabrication des mesures de longueur à l'usage des ouvriers (Paris, 1795), p. 1, listed six measures that were of special interest to laborers: decameter, double meter, meter, half meter, double decimeter, and decimeter. On related matters see L'Agence Temporaire des Poids et Mesures, Paris, Aux Citoyens rédacteurs de la feuille du cultivateur (Paris, 1795), pp. 7 ff. and Leon Baptiste Auguste Barny de Romanet, Traité historique des

<u>poids et mesures et de la vérification depuis Charlemagne jusqu'à nos jours</u> (Paris, 1863), pp. 53 ff. A listing of these new units compared with their contemporary pre-metric and English equivalents is in Patrick Kelly, <u>Metrology: or, An Exposition of Weights and Measures, Chiefly Those of Great Britain and France</u> (London, 1816), pp. 21 ff.

3. L'Agence Temporaire des Poids et Mesures, Paris, <u>Tables de comparaison entre les mesures, anciennes et celles qui les remplacent dans le nouveau système métrique, avec leur explication et leur usage</u> (Paris, 1796), pp. 12 ff. Also see C.L. Aubry, <u>Le Système des nouvelles mesures de la République française</u> (Paris, 1797), pp. 11 ff. Between 1795 and 1797 many suggestions were made by various sectors of French society to increase the number of acceptable units in order to cover more of the customary units of the ancient regime. As a result of scientific, technological, commercial, business, agricultural, labor, medical, pharmaceutical, and other interests, many units were added by 1798 that eventually were eliminated by 1800 as either impractical or superfluous. Among them, including some found earlier in the law of 1795, were the myriameter, half kilometer, double hectometer, half hectometer, double dekameter, half dekameter, double meter, half meter, and double decimeter in linear measurements; the myriare, kilare, half kilare, double hectare, half hectare, double dekare, half dekare, double are, half are, double deciare, deciare, centiare, and milliare among the superficial measures; the myriastere, kilostere, half kilostere, double hectostere, hectostere, half hectostere, double

dekastere, dekastere, half dekastere, double stere, half stere, double decistere, decistere, centistere, and millistere of volume measures; the myrialiter, half kiloliter, double hectoliter, half hectoliter, double dekaliter, half dekaliter, double liter, half liter, and double deciliter for liquid and dry capacity measures; and the myriagram, half kilogram, double hectogram, hectogram, half hectogram, double dekagram, half dekagram, double gram, half gram, double decigram, decigram, centigram, and millionigram among the weights.

4. C.F. Delandes de Bagneux, <u>Tables des rapports des mesures du système métrique, déduites de la grandeur de la terre, et des anciennes mesures du département de la Drôme, avec leurs prix comparatifs, et un tableau général du nouveau système</u> (Lyon, 1804), pp. 56-57. Most of these complaints would be uttered verbatim by English anti-metric spokesmen.

5. The exact figures were 6,075.900069 fathoms for the base at Melun and 6,006.247848 fathoms for the base at Perpignan. The terminals of the base line were marked by a solidly constructed, cut-masonry, monument, the foundation of which rested on rock. The exact point was marked by a copper cylinder inserted in the monument, the terminal point being the center of several concentric circles drawn on the surface of the cylinder. This cylinder was placed in a square depression of the monument and was covered with a layer of lead, upon which was placed the fitting pyramidal capstone, projecting a little above the surface of the ground. The ground was then paved, and the whole surrounded by cut stone pillars a few feet in height, arranged in circular form.

6. C.F. Delandes de Bagneux, Tables les rapports des mesures, p. 6. In the measurement procedures for the meridian, it was decided that the measuring rods would be constructed of the most durable metal, and platinum was chosen. They were each 12 feet in length, 6 lines wide, and almost 1 line thick. These rods were numbered I, II, III, and IV, with rod number I, upon which the others were sized, receiving the name "module."
7. Considering the earth as a spheroid of revolution, the committee determined the length of a quadrant of the meridian to be 5,130,740 semi-modules, the ten-millionth part being 0513.074 semi-module. Fixing the semi-module as the equivalent of 864 lines, they made the meter equal to 443.296 lines. By comparing an English standard made by Troughton, this produced a figure for the new platinum meter equal to 29.3781 English inches, and one of the iron meters equal to 39.3795 English inches.
8. Other European states had similar make-shift systems. In various German kingdoms, for example, the following units were used: the Decimallinie, varying from 0.0028 m at Dresden and Leipzig to 0.0047 m at Hannover; the Decimalfuss of 10 Decimalzoll, varying from 0.282 m at Leipzig to 0.467 m at Hannover; the Decimalzoll of 10 Decimallinien, varying from 0.0282 m at Leipzig to 0.0467 m at Hannover; the Neuzoll of 1 centimeter, used everywhere after 1868; the Prime of 10 Sekunden (0.0026 m), used in Berlin from 1816 to 1872; the Sekunde of 0.00026 m, used in Berlin during the same period; and the Zehntelruthe of

Chapter 5

Frankfurt, containing 10 Zoll of 10 Linien each (0.285 m).

Chapter 6

1. Among linear units these included the foot, inch, perch, and yard. Allowable weights were the cheese clove, several hundredweights, tower and troy ounces and pennyweights, avoirdupois, tower, and troy pounds, wool stone, hay and straw trusses, and the cheese wey. The acre was the only superficial measure ever defined by law. For capacity measures, the largest group, one finds the ale, beer, eel, herring, soap, vinegar, and wine barrels, coal, grain, and salt bushels, salmon butt, coal cartload, coal chalders, herring cran, ale, beer, butter, and soap firkins and kilderkins, grain and wine gallons, gill, wine hogshead, coal keels, peck, pint, wine pipe, butter pot, grain pottle, ale and grain quarts, quarter, wine runlet, coal and wool sacks, wine tertian, wine tun, and the coal wagonload.
2. Some of the general developments in these nineteenth-century parliamentary initiatives may be found in Henry James Chaney, _Our Weights and Measures: A Practical Treatise on the Standard Weights and Measures in Use in the British Empire with Some Account of the Metric System_ (London, 1897), pp. 1-34; H.W. Chisholm, _On the Science of Weighing and Measuring, and the Standards of Weight and Measure_ (London, 1877), pp. 66-96; "Weights and Measures," in _The American Cyclopaedia: A Popular Dictionary of General Knowledge_, ed. George Ripley and Charles A. Dana (New York, 1876), vol. 16, pp. 537-544;

Thomas C. Mendenhall, "Fundamental Units of Measure," in Annual Report of the Board of Regents of the Smithsonian Institution...to July, 1893 (Washington, D.C., 1894), pp. 135-149; William Harkness, "The Progress of Science as Exemplified in the Art of Weighing and Measuring," in Smithsonian: Miscellaneous Collections (Washington, D.C., 1888), vol. 33, pp. 597-633; "Third Report of the Commissioners Appointed by His Majesty to Consider the Subject of Weights and Measures," in Parliamentary Papers, Great Britain: Report from Committees (London, 1821), vol. 4, pp. 1-6; J.H. Alexander, An Inquiry into the English System of Weights and Measures (Oxford, 1857), pp. 28-37, 85-89; Thomas Young, "On Weights and Measures," in Miscellaneous Works (London, 1855), pp. 427-435; S.C. Walker, "Report on the Weights and Measures of Great Britain," in Report of the Managers of the Franklin Institute (Philadelphia, 1834), pp. 33-48; Henry Lyons, The Royal Society: 1660-1940 (New York, 1968), pp. 173-175, 204-205, 220-227; W.S.B. Woolhouse, Measures, Weights, & Moneys of All Nations (London, 1890), pp. 1-28; and F. Baily, "Report of the New Standard Scale of This Society," in Memoirs of the Royal Astronomical Society (London, 1836), vol. 9, pp. 35-45.

3. This act was to take effect on May 1, 1825, but in March of that year an act of George IV deferred its operation due to organizational difficulties until January 1, 1826.

4. In the metric system, of course, there were only two such standards—the definitive meter and kilogram. The first governmental use of the word

"imperial" to designate this new system of weights and measures occurred in the third report of the commissioners in 1821.

5. Other enactments in the 1824-1878 period that mentioned weights and measures in relation to certain trades and products were the Bread Acts of the 1820s and 1830s, the London Coal Act of 1831, the Beerhouse Act of 1834, the Railways Clauses Consolidation Act of 1845, the Hay and Straw Act of 1856, the Sale of Gas Act of 1859, the Refreshment Houses Act of 1860, the Hop (Prevention of Frauds) Act of 1866, the Sea Fisheries Act of 1868, the Judicature Act of 1873, and the Sale of Food and Drugs Act of 1875.

Similar enactments, thereafter, were the Spirits Act of 1880, the Corn Returns Act of 1882, the Coal Mines Regulation Act of 1887, the London Coal Duties Abolition Act of 1889, the Herring Fisheries (Scotland) Act of 1889, the Sale of Food and Drugs Act of 1889, the Factory and Workshop Act of 1901, the Finance Acts of 1907 and 1908, the Licensing (Consolidation) Act of 1910, the Herring Fishery (Branding) Act of 1913, the Ministry of Agriculture and Fisheries Act of 1919, the Seeds Act of 1920, the Licensing Act of 1921, the Judicature Act of 1925, the London County Council (General Powers) Act of 1928, the Food and Drugs (Adulteration) Act of 1928, the Measuring Instruments (Liquid Fuel and Lubricating Oil) Regulations of 1929, the Factories Act of 1937, the Food and Drugs Act of 1938, the Statutory Instruments Act of 1946, the Gas Act of 1948, the Pre-packed Food (Weights and Measures: Marking) Order of 1950, the Food and Drugs Act of 1955, the Sale of Food (Weights and Measures: Bacon and Ham) Regulations of 1956, and the Pre-packed Food (Weights and Measures: Marking) Regulations of 1957.

6. The conflagration was due to the burning of the "tallies" or sticks on which treasury accounts had been kept by means of notches at the Exchequer. See Henry Kater, "An Account of the Construction and Adjustment of the New Standards of Weights and Measures of the United Kingdom of Great Britain and Ireland," in Philosophical Transactions (London, 1826), vol. 116, pp. 46-48, for a listing of all the contemporary legal imperial standards kept in Edinburgh, Dublin, the Exchequer at Westminster, and the Guildhall in London.

7. The standards that were constructed were actually kept in a special air- and temperature-conditioned vault under the charge of the Standards Department of the Board of Trade.

8. It must be pointed out, however, that even though the Weights and Measures Act of 1963 retained the 1855 yard and pound as the United Kingdom primary standards for commercial use, the international yard was defined as exactly 0.9144 meter, and the meter was defined in terms of the wave length of light radiated by krypton 86. The international pound was defined as the exact weight of 0.45359237 of the international prototype kilogram.

9. The bar was marked: Copper 16 oz. Tin 2 1/2 Zinc 1 Mr. Baily´s Metal No. 1 STANDARD YARD at 62°.00 Faht. Cast in 1845 Troughton & Simms, LONDON.

10. See Figure 15 for the imperial apothecaries measures constructed after the issuance of this act.

11. W. Roger Breed in The Weights and Measures Act: 1963 (London, 1964), p. 18, asks whether more efficiency would be achieved if the inspection of weights and measures were placed under the central authority, as in France, rather than under numerous local authorities.

12. It was understood at the time of this act´s passage that a few courts leet continued to appoint inspectors, but provision was made by the Weights and Measures (Purchase) Act of 1892 for the purchase by county and borough councils of franchises of weights and measures with a view to their future extinction. Also, nothing in the 1878 enactment negated the rights of the Lord Mayor of London as official gauger of wines, oil, honey, and other gaugeable liquors imported or brought into the port of the city of London.

13. Most of the methods to be followed in carrying out the verification of weights, measures, and weighing and measuring instruments are outlined in detail in the Weights and Measures Regulations of 1907, the Measuring Instruments (Liquid Fuel and Lubricating Oil) Regulations of 1929, the Weights and Measures (Sand and Ballast) Regulations of 1938, the Weights and Measures (Amendment No. 4) Regulations of 1942, and the Weights and Measures (Bell Weights: Verification and Stamping) Regulations of 1952. Inspectors are also responsible for the verification and stamping of weighing instruments used for trade by the Weights and Measures Act of 1889; of leather measuring instruments by

the Weights and Measures (Leather Measurement) Regulations of 1921; of certain instruments used for measuring liquid fuel and lubricating oil by the Measuring Instruments (Liquid Fuel and Lubricating Oil) Regulations of 1929; and of receptacles used or intended to be used for purposes of trade in measuring sand or ballast by the cubic yard by the Weights and Measures (Sand and Ballast) Regulations of 1938. For inspectors´ duties in relation to the verification of weights, etc. used in mines and factories, see the Coal Mines Regulations Act of 1887 and the Factories Act of 1937. Additional information may be gathered from the Sale of Food (Weights and Measures) Act of 1926 and the Weights and Measures Act of 1936.

14. Besides this type of record, the Weights and Measures Regulations of 1907 instructed local authorities to request from each of their inspectors an annual report concerning the inspector´s examination and verification functions over the past year. Copies of these reports must be sent by each local authority to the Board of Trade not later than September 30 in any year. To insure the accuracy of the annual report, each inspector must keep an "Inspection Book," a "Working Diary," and a "Certificate of Verification" book.

15. Just before the passage of this act, W. and T. Avery, Suggestions for the Amendment of the Law Relating to Weights & Measures (London, 1888), pp. 10 ff., saw English metrological law as defective and incomplete since it ignored the advances made by the French and lacked an authoritative general administrative method of inspection. They

believed that the German and French systems (the Kaiserlichen Normal-Aichungs-Kommission and the Ministry of Commerce and Agriculture) to be superior. Also, technical literature abounded on the continent. In England, the Averys labored for tighter control by the Standards Department over weights and measures personnel and over scales and weighing machines, for limiting the arbitrary powers of inspectors, for more professionally trained inspectors, for limiting local control, and for governmental certification in all aspects of metrological control. The law of 1889 helped considerably.

Chapter 7

1. Taylor states (p. 48) that Baily once remarked that the standard yard kept at the Exchequer before the work of the Imperial Commissions was no scientific measure of accuracy at all. "A common kitchen poker, filed at the ends by the most bungling workman, would have made as good a standard." He was more impressed with the imperial standards, but he believed that they had to be altered to conform to a decimal calibration.
2. Jessop's reason for selecting the number 64 was that it is the sixth power of 2; thus the pendulum could be divided in 2 equal parts 6 times successively.
3. See Commission Internationale du Mètre, Réunions Générales de 1872: Procès-Verbaux (Paris, 1872), pp. 163-171, 185-209; Comité International des Poids et Mesures, Procès-Verbaux des séances de 1878

(Paris, 1879), pp. 239-271; Ibid, Procès-Verbaux des séances: Session de 1909 (Paris, 1909), vol. 5, pp. 121-141; Ibid, Procès-Verbaux des séances: Session de 1911 (Paris, 1911), vol. 6, pp. 193-213; and Comptes rendus des séances de la Septième Conference Générale des Poids et Mesures réunie à Paris en 1927 (Paris, 1927), pp. 83-121.

4. The French government donated the Pavillon de Breteuil, including a large tract of land situated on the bank of the Seine near Sèvres, as the site for the International Bureau. The laboratory dates from 1878; other buildings came later. See H.T. Wade, "International Bureau of Weights and Measures," in Scientific American, 98 (February 8, 1908), pp. 93-94; C.E. Guillaume, "International Bureau of Weights and Measures," in Scientific American Supplement, 86 (September 21, 1918), pp. 186-187; Henri Moreau, "The Genesis of the Metric System and the Work of the International Bureau of Weights and Measures," in Journal of Chemical Education, 30, 2 (1953), pp. 3-20; "Labors of the International Bureau of Weights and Measures," in Scientific American Supplement, 29 (January-June), 1890), pp. 11821-11823; and René Benoit, "Modification de la législation française relative aux unités fondamentales du système métrique," in Comité International des Poids et Mesures, Procès-Verbaux des séances (Paris, 1905), vol. 3, pp. 139-174.

5. See Ronald Edward Zupko, "Worldwide Dissemination of the Metric System During the 19th and 20th Centuries," in Metric System Guide Bulletin (Neenah, Wisconsin, 1974), vol. 2, no. 2, pp. 14-25 for a complete

listing of the nations that adopted the metric system from the early 1800s to the 1970s, together with the names of the laws involved and the dates of introduction and adoption.

Also see the following works for additional discussions of metric developments during this era: "An Act to Render Compulsory the Use of the Metric System of Weights and Measures," in Parliamentary Papers, Great Britain: Sessional Papers (London, 1868, 1871, 1904); "An Act to Render Permissive the Use of the Metric System of Weights and Measures," in Parliamentary Papers, Great Britain: Sessional Papers (London, 1897); H. Barrell, "A Short History of Measurement Standards at the National Physical Laboratory," in Contemporary Physics, 9 (1968), pp. 205-226; A. Biremhaut, "Les deux Déterminations de l´unité de masse du système métrique," in Revue d´histoire des sciences, 12 (1959), pp. 25-54; William K. Burton, Measuring Systems and Standards Organizations (New York, 1972); Fifth Report of the Commissioners Appointed to Inquire into the Condition of the Exchequer (Now Board of Trade) Standards. Parliamentary Papers, Great Britain: Sessional Papers (London, 1893); Final Report of the Committee on Commercial and Industrial Policy After the War. Parliamentary Papers, Great Britain: Sessional Papers (London, 1918); First Report of the Commissioners Appointed to Inquire into the Condition of the Exchequer (Now Board of Trade) Standards. Parliamentary Papers, Great Britain: Sessional Papers (London, 1869); Charles Edouard Guillaume, La Convention du Mètre et le Bureau International des Poids et Mesures (Paris, 1902), La

Création du Bureau International des Poids et Mesures et son oeuvre (Paris, 1927), Les récents Progrès du système métrique (Paris, 1907), and Unités et étalons (Paris, 1894); Alexis Guillemot, Renseignements sur le service, la vérification et la fabrication des poids et mesures (Châlons-sur-Marne, 1902); D. Isaachsen, La Création du Bureau International des Poids et Mesures et son oeuvre (Paris, 1927); Gustav Karsten, Die internationale General-Konferenz für Maass und Gewicht...1889 (Kiel, 1890); "Minutes of Evidence Taken Before the Select Committee on the Bill to Amend and Render More Effectual Two Acts of the Fifth and Sixth Years of the Reign of His Majesty King George the Fourth Relating to Weights and Measures," in Parliamentary Papers, Great Britain: Reports from Committees of the House of Lords, 18 (1835), Part 1; National Physical Laboratory, Teddington, England, Balances, Weights, and Precise Laboratory Weighing (London, 1962); "Report from the Select Committee on the Weights and Measures Act: Together with the Minutes of Evidence," in Parliamentary Papers, Great Britain: Reports from Committees of the House of Lords, 18 (1835), pp. 1-60; Report from the Select Committee on Weights and Measures; together with the Proceedings of the Committee, etc. Parliamentary Papers, Great Britain: Sessional Papers (London, 1862 and 1895); Report from the Select Committee on Weights and Measures (Metric System) Bill. Parliamentary Papers, Great Britain: Sessional Papers (London, 1904); "Report of the Commissioners Appointed to Consider the Steps to be Taken for Restoration of the Standards of Weights and

Measures," in Parliamentary Papers, Great Britain: Reports from Commissioners, 25 (1842), pp. 1-106; "Report of the Committee Appointed to Superintend the Construction of the New Parliamentary Standards of Length and Weight," in Parliamentary Papers, Great Britain: Reports from Committees of the House of Lords, 19 (1854), pp. 1-23; Second Report of the Commissioners Appointed to Inquire into the Condition of the Exchequer (Now Board of Trade) Standards. etc. Parliamentary Papers, Great Britain: Sessional Papers (London, 1869); "The Select Committee Appointed to Consider the Several Reports Which Have Been Laid Before This House Relating to Weights and Measures," in Parliamentary Papers, Great Britain: Reports from Committees of the House of Lords, 4 (1821), pp. 1-7; J. Terrien, Le Changement de la définition du mètre et le rôle du Bureau International des Poids et Mesures (Paris, 1960); Charles Testut, Mémento du pesage; les instruments de pesage; leur histoire à travers les âges (Paris, 1946); Third Report of the Commissioners Appointed to Inquire into the Condition of the Exchequer (Now Board of Trade) Standards. Parliamentary Papers, Great Britain: Sessional Papers (London, 1869); and C. Wolf, "Recherches historiques sur les étalons de poids et mesures de l´Observatoire et les appareils qui ont servi à les construire," in Annales de l´Observatoire de Paris, 17 (1883), pp. 30-60.

6. The countries, with their dates of introduction and final adoption, were: Italy (1803-63), Belgium (1816), Luxembourg (1816-20),

Netherlands (1816-32), Greece (1836-1959), Senegal (1840), Algeria (1843), Chile (1848-65), Cuba (1849-82), Dominican Republic (1849-1955), Philippines (1849-1917), and Spain (1849-71).

7. Portugal (1851-72), Colombia (1853-54), Panama (1853-1916), Monaco (1854), Ecuador (1856-71), Mexico (1857-96), Venezuela (1857-1914), Brazil (1862-72), Peru (1862-69), Uruguay (1862-94), Argentina (1863-87), Romania (1864-84), Bolivia (1868-71), Germany (1868-72), Switzerland (1869-77), Austria (1871-76), Czechoslovakia (1871-76), India (1871-1956), Yugoslavia (1873-1912), Hungary (1874-76), Liechtenstein (1875-76), Norway (1875-82), Sweden (1878-89), Chad (1884-1907), Gabon (1884-1907), Ivory Coast (1884-1907), Mauritania (1884-1907), Sudan (1884-1907), Upper Volta (1884-1907), Costa Rica (1885-1912), El Salvador (1886-1912), Finland (1886-92), Tunisia (1886-95), Turkey (1886-1933), Bulgaria (1888-92), Thailand (1889-1936), Dahomey (1890-91), United Arab Republic (1891-1939), Japan (1893-1959), Nicaragua (1893-1912), Cameroon (1894-1924), Guatemala (1894-1912), Honduras (1897-1912), Somali Republic (1898), Paraguay (1899), and U.S.S.R. (1900-27).

8. Guinea (1901-06), Denmark (1907-12), Iceland (1907), San Marino (1907), Republic of Congo (Brazzaville) (1910), Democratic Republic of Congo (Leopoldville) (1910), Malta (1910-21), Viet-Nam (1911), Cambodia (1914), Republic of China (1914-30), Poland (1919), Haiti (1920-22), Morocco (1923), Indonesia (1923-38), Togo (1924), Afghanistan (1926), Libya (1927), Iraq (1931-60), Iran (1933-49), Lebanon (1935), Syria

(1935), Israel (1947-54), and Korea (1949).

9. Albania (1951), Jordan (1953-54), Kuwait (1961-64), Saudi Arabia (1962-64), Nepal (1963-71), Ethiopia (1967), Kenya (1967), Pakistan (1967), South Africa (1967), Tanzania (1967), and Uganda (1967).

10. The same problems plagued American metrication efforts during this period and beyond. After some encouraging moves in the late 1700s and early 1800s, the question of metrological reform was pushed far into the background. The only intense interest prior to the Civil War occurred in 1816, when President James Madison suggested that a decimal system of weights and measures be adopted. The Senate responded by issuing a resolution in March, 1817, that instructed John Quincy Adams, the Secretary of State, to prepare a report on reform. Several plans were advanced but none espoused metrication, and the issue was dropped. In 1843, the Committee on Commerce reported negatively on metrics once again. After the War, interest resurfaced when Congress reluctantly passed the Metric System Act of 1866 that officially recognized the use of metric weights and measures in commercial transactions. In a joint resolution approved one day before the Act, Congress directed the Secretary of the Treasury to furnish each state with a set of metric standards. Unfortunately the Act did not solve the metric dilemma; it only went halfway by not eliminating the customary English system. All that the Act did was to recognize the legitimacy of contracts in which dimensions or sizes were expressed in metric terms. For the remainder of this century, and for the first seven decades of the twentieth

century, there were dozens of bills and hundreds of committee reports and recommendations before Congress that contained thousands of provisions aimed at establishing varying degrees of metric usage. Everyone of them failed in their objectives—even a Bill of 1902 that, prior to the vote, was almost assured of passage failed due to a long delay between the adjournment of one Congress and the inception of its successor.

11. See part 5, chapter 6 of Leone Levi, The History of British Commerce and of the Economic Progress of the British Nation: 1763-1878 (London, 1880) and Report of the International Conference on Weights, Measures, and Coins, Held in Paris, June 1867 (Communicated to Lord Stanley by Professor Leone Levi), Parliamentary Papers, Great Britain: Sessional Papers (London, 1868).

12. The official title was An Act to Render Permissive the Use of the Metric System of Weights and Measures, 29 July 1864.

13. The International Association had ceased most of its functions before this, and the British branch no longer concerned itself with adoption. The reasons for the latter's disinterest may be due to the fact that it considered the legislation of 1864 to be a "paper" victory, or it may have thought that the metric system would triumph eventually anyway, or that its basic work was accomplished and that the responsibility was now the burden of the British Association for the Advancement of Science, a far more vocal and financially solvent organization. Whatever the ultimate reason, future metric programs drew upon the

expertise and endorsement of other groups.

14. Major publications treating the issues raised in this controversy are Herbert Spencer, Against the Metric System (London, 1904), pp. 3-24; Thomas C. Mendenhall, The Metric System (New York, 1896), pp. 2-13; George Eastburn, The Metric System (New York, 1892), pp. 3-21; Frederick A.P. Barnard, The Metric System of Weights and Measures (New York, 1872), pp. 7-15, 71-113, 179-189; Alfred Watkins, Must We Trade in Tenths? (Hereford, 1919), pp. 1-8; J. Pickering Putnam, The Metric System of Weights and Measures (Boston, 1877), pp. 17-55; Frederick A. Halsey, The Metric Fallacy (New York, 1920), pp. 77-87; F. Mollwo Perkin, The Metric and British Systems of Weights, Measures and Coinage (New York, 1907), pp. 9-11; John R. Townsend, "Metric Versus English Systems," in Systems of Units: National and International Aspects, ed. Carl F. Kayan (Washington, D.C., 1959), pp. 193-200; W. H. Wagstaff, The Metric System of Weights and Measures Compared with the Imperial System (London, 1896), pp. 14-15; John William Nystrom, On the French System of Weights and Measures, with Objections to Its Adoption Among English-Speaking Nations (Philadelphia, 1876), pp. 5-15; Alexander Craig Aitken, The Case Against Decimalization (Edinburgh, 1962), pp. 5-22; Robert Henry Thurston, Conversion Tables of Metric and British or United States Weights and Measures (New York, 1883), pp. 5-20; "British and Metric Measures," in Scientific American Supplement, 19 (January-June, 1885), p. 7661; "An Adverse View of the Metric System," in Ibid, 55 (January-June, 1903), pp. 22579-22580; John Perry, The

<u>Story of Standards</u> (New York, 1955), pp. 73-92; British Association for the Advancement of Science, <u>Decimal Coinage and the Metric System: Should Britain Change?</u> (London, 1960), pp. 28-101; Alexander Siemens, "The Metric System," in <u>Scientific American Supplement</u> (New York, 1914), 78, pp. 6-7; Hugh Oakeley Arnold-Forster, <u>The Coming of the Kilogram</u> (London, 1905), pp. 52-59, 86-89, 116-135; Julia Emily Johnsen, comp., <u>Metric System</u> (New York, 1926), pp. 7-19; and C. McL. McHardy, <u>Proposal for Decimal Coinage and Metric Weights and Measures</u> (London, 1902), pp. 1-10 and Appendix F.

Other articles of importance for presenting detailed arguments both for and against the metric system are "Abstract of Report Given by Consul Boyle of Liverpool Showing Great Saving from Decimalizing One Unit—the Half-Hundredweight," in <u>Science</u>, 19 (1904), pp. 119-120; "The Adoption of the Metric System," in <u>Nature</u>, 99 (August 30, 1917), pp. 526-527; Harry Allcock, "The Decimal System of Coinage, Weights, and Measures," in <u>Surveyor and Municipal and County Engineer</u>, 51 (1917), pp. 388-389, 415-416; Charles Atherton, "On Decimalising British Measures, Weights, and Coins," in <u>Journal of the Society of Arts</u>, 8 (1859-1860), pp. 591-593; Llewellyn B. Atkinson, "The Metric System," in <u>Institution of Electrical Engineers of Great Britain</u>, 56 (February 18, 1918), pp. 121-125; Auguste Aurès, "Note sur le système métrique anglais," in <u>Revue archéologique</u>, 15 (1867), p. 444; Aldred F. Barker, "Adoption of the Metric System," in <u>Decimal Educator</u>, 1 (1918-1919), pp. 45-48; William Barlow, "An Account of the Analogy betwixt English

Weights and Measures of Capacity," in Philosophical Transactions, 41 (1740), pp. 457-459; Frederick A.P. Barnard, "Metric System," in New York University Regents´ Report, 85 (1872), pp. 585-691; Eugene C. Bingham, "Progress in Metric Standardization," in Science, 55 (March 3, 1922), pp. 232-233; John A. Brashear, "Evolution of Standard Measurements," in American Manufacturer and Iron World, March 29, 1900, pp. 256-258; A. Briggs, "For Introduction of the Metric System," in Franklin Institute Journal, 100 (1875), pp. 145-152; "British Weights and Measures and the Metric System," in Edinburgh Review, 212 (October, 1910), pp. 426-449; S. Brown, "On the Metric System and Its Proposed Adoption in Great Britain," in Society of Arts Journal, 12 (January 29, 1864), pp. 162-170; J.V. Collins, "Metric Weights and Measures," in Education, 15 (1894), pp. 229-235; "The Decimal Association: An Historical Sketch," in Decimal Educator, 1 (1918-1919), p. 4; "Discussion on the Metric System," in Franklin Institute Journal, 153 (1902), pp. 401-418; J. Emerson Dowson, "Decimal Coinage, Weights, and Measures," in Journal of the Society of Arts, 39 (1890-1891), pp. 201-223; N.F. Dupuis, "The Metric System," in Queen´s Quarterly, 14 (1907), pp. 163-172; C.W. Eliot, "Metric System of Weights and Measures," in Nature, 2 (June 8, 1866), p. 731; "England and the Metric System," in Scientific American Supplement, 72 (July 1, 1911), p 59; Persifor Frazer, "The Metric System of Weights and Measures," in Franklin Institute Journal, 154 (1902), pp. 107-111; David Gill, "The Science of Measurement: Its Application to Astronomical Research," in

Scientific American Supplement, 64 (August 24, 1907), pp. 126-128; H. McDonald Glasgow, "Metric System in England," in Scale Journal, 3 (1917), pp. 11-14; J. Howard Gore, "Metric System and International Commerce," in Forum, 31 (1901), pp. 739-744; E. Sherman Gould, "An Adverse View of the Metric System," in Scientific American Supplement, 55 (January-June, 1903), pp. 22579-22580; Frederick A. Halsey, et al., "The Metric System Fallacy," in Cassier's Magazine, 30 (1906), pp. 36-57; W.W. Hardwicke, "Currency and Weights and Measures: The Case for Reform," in Empire Review, December 1915, pp. 503-510; Joseph Hartigan, "Metric System," in Scientific American Supplement, 83 (March 17, 1917), pp. 175-176; C. Doris Hellman, "Jefferson's Efforts towards the Decimalization of United States Weights and Measures," in Isis, 16 (1931), pp. 266-272; Walter Renton Ingalls, "Shall Great Britain and America Adopt the Metric System?," in Journal of the Society of Arts, 65 (1916-1917), pp. 604-610; "International Committee of Weights and Measures," in Nature, 87 (August 24, 1911), pp. 251-252; "International Conference on the Meter," in Scientific American Supplement, 77 (January 3, 1914), p. 67; Keith Gordon Irwin, "Fathoms and Feet, Acres and Tons: An Appraisal," in The Scientific Monthly, 72 (1951), pp. 11-12; George Frederick Kunz, "The International Language of Weights and Measures," in Scientific Monthly, 4 (March, 1917), pp. 215-219; Charles MacDonald, "Metric System of Weights and Measures," in Royal Philosophical Society of Glasgow: Proceedings, 41 (1910), pp. 65-83; W.W. McFarland, "The Metric System Fallacy," in Cassier's Magazine, 29

(1906), pp. 293-299; Herbert McLeod, "Notes on the History of the Metrical Measures and Weights," in Nature, 69 (1904), pp. 425-427; "The Metric Controversy," in Scientific American Supplement, 58 (July-December, 1904), p. 24108; "The Metric System," in Scientific American Supplement, 49 (January-June, 1900), pp. 20197-20198; "Metric System and Decimal Coinage," in Nature, 96 (January 20, 1916), pp. 568-570; "Metric System and the Decimal," in Westminster Review, 131 (1889), pp. 280-285; "Metric System in Great Britain," in Science, 16 (1902), pp. 595-596; "Metric System in the British Colonies," in Nature, 74 (October 18, 1906), pp. 614-615; F.J. Miller, "Metric System of Weights and Measures," in Science, 37 (July 20, 1913), pp. 941-942; George Moores, "Inch vs. the Meter," in Cassier´s Magazine, 27 (1904), pp. 157-160; H.A. Newton, "The Metric System of Weights and Measures, with Tables," in Annual Report of the Board of Regents of the Smithsonian Institution...for the Year 1865, Washington, D.C., 1872, pp. 465-485; "New Units in the Metric System, Legally Adopted in France," in Scientific American Monthly, 2 (October, 1920), p. 152; "Plea for Metric System," in Science, 23 (January 12, 1906), pp. 73-74; "Position of the Metric System," in Nature, 79 (February 25, 1909), p. 501; "The Present Position of the Metric System in the British Empire and the United States," in Decimal Educator, 1 (1918-1919), pp. 26-28; J.P. Putnam, "Advantages of the Metric System," in Franklin Institute Journal, 100 (1875), pp. 145-152; W.F. Quinby, "The Metric System," in International Standard, 1 (1883-1884), pp. 18-20; "Recent Progress in the Metric System," in Nature, 93 (July 9, 1914), p. 483; "Relation of the Metric System to the Triumph of German over Anglo-Saxon Efficiency," in Current Opinion, 60 (March, 1916), p. 186; "Report on the Metric System," in Scientific American Supplement, 53 (May 10, 1902), pp. 36-38; "Review of Bill Establishing Metric System in Great Britain," in Nature, 3 (April 6, 1871), pp. 448-449; Howard Richard, "The Metric Campaign," in Science, 55 (May 12, 1922), pp. 515-516; H.W. Richardson, "The Metric System," in Harper´s Monthly, 81 (1890), pp. 509-513; William A. Rogers, "On the Present State of the Question of Standards of Length," in Proceedings of the American Academy of Arts and Sciences, 15 (1879-1880), pp. 273-312; Alfred Sang, "The Metric System," in Scientific American, 94 (1906), pp. 454-455; W.H. Seaman, "The Bill in the British Parliament," in Science, 21 (1905), p. 72; Coleman Sellers, "Metric System," in Cassier´s Magazine, 17 (1900), pp. 365-377; Napier Shaw, "Units and Unity," in Nature, 101 (1918), pp. 326-328; A. Sonnenschein, "The Metric System," in Journal of the Society of Arts, 51 (1902-1903), pp. 170-178; A.C. Stubbs, "A Case for the Adoption of the Metric System (and Decimal Coinage) by Great Britain," in Institution of Electrical Engineers of Great Britain, 56 (February 18, 1918), pp. 125-135; "Text of Bill to Establish Metric System in British Isles," in Nature, 3 (1871), pp. 448-449; J.B. Thompson, "The Metric System," in New York University Regents´ Report, 86 (1873), pp. 659-680; Charles A.L. Totten, "Why Anglo-Saxon Metrology Should Not Be Abandoned," in International Standard, 1 (1883-1884), pp.

371-375; Spencer M. Vawter and Ralph E. Deforest, "The International Metric System and Medicine," in Journal of the American Medical Association, 218 (November 1, 1971), pp. 723-726; Charles Moore Watson, "Some Objections to the Compulsory Introduction of the Metric System," in Journal of the Society of Arts, 55 (1906-1907), pp. 50-64; N.J. Werner, "Metric System in Typography," in Measurement, 6 (April, 1931), pp. 4-7; W.P. White, "Powers of Ten," in Science, 35 (January 5, 1912), pp. 38-40; Leo Wiener, "Origin of the Metric System," in Nation, 84 (1907), pp. 103-104; A.B. Willmott, "The Metric System," in Engineering and Mining Journal, 79 (1905), pp. 766-767; and James Yates, "Narrative of the Origin and Formation of the International Association for Obtaining a Uniform Decimal System of Measures, Weights and Coins," in Metric System Pamphlets, 1 (1856), No. 11.

In addition to this rich journal lode of information on the metric controversy, readers should consult the following monographs: Harry Allock, Commentary upon the Findings of Lord Balfour of Burleigh´s Committee with Regard to Decimal Coinage and the Metric System (London, 1918), The Decimal System of Coinage, Weights and Measures (London, 1917), Great Britain´s Interest in the Metric System of Weights and Measures (London, 1918), and Industrial Reconstruction and the Metric System (London, 1919); George Frederick Allwood, Appeal Cases under the Weights and Measures Acts (London, 1906); J.S.H. Aslit, Decimal Coinage: With a Proposal for Decimalizing Our Weights and Measures (London, 1854); Alfred Barker, Unification of Weights and Measures in the Textile Industries: The Possibilities of the Metric System (London, n.d.); E.C. Barton and George Morres, For and Against the Metric System (London, 1916); D. Beach and E.A. Gibbens, The Metric System and Interchange of Weights and Measures (New York, 1880); British Weights & Measures Association, London, Second Annual Report (London, 1906); Frederick Brooks, Metric System of Weights and Measures (Boston, 1875); Samuel Brown, On the Metric System of Weights and Measures, and Its Proposed Adoption in This Country (London, 1864); George W. Colles, The Metric Versus the Duodecimal System (Boston, 1896); Conjoint Board of Scientific Societies, London, Report on Compulsory Adoption of the Metric System in the United Kingdom Submitted by the Metric Committee, etc. (London, 1920); Edward Franklin Cox, A History of the Metric System of Weights and Measures, with Emphasis on Campaigns for Its Adoption in Great Britain and in the United States Prior to 1914 (University Microfilms, Ann Arbor, 1958); Benjamin F. Craig, Weights and Measures According to the Decimal System (New York, 1867); Samuel S. Dale, The Mendenhall Conspiracy to Discredit English Weights and Measures (Boston, 1927) and Testimony Against the Metric System, etc. (Boston, 1907); Charles Davies, The Metric System (New York, 1867); Decimal Association, London, Decimal Coinage: The Existing British Coinage Re-arranged on the Decimal Scale (London, n.d.), Extracts from the Evidence Given by Lord Kelvin, before the Committee on Coinage, Weights and Measures...24 April, 1902 (London, 1903), Is a Duodecimal System Possible? (London, n.d.),

Reprint from The Parliamentary Debates, 23 February, 1904 (London, 1904), and *Statement Submitted to the Dominions Royal Commission by the Decimal Association on 4 October, 1912* (London, 1913); Fred R. Drake, *The Metric System* (Washington, D.C., 1914); Aubrey Drury, *The Metric Advance: The Move to Standardize and Decimalize the Yard-Quart-Pound on the World Metric Basis* (Washington, D.C., 1926); A.J. Ede, *Advantages of the Metric System* (London, 1972); J.W. Evans, *British Weights and Measures Considered from a Practical Standpoint: A Plea for Their Retention in Preference to the Metric System* (Sydney, South Wales, 1904); Frank Perks Fellows, *On the Impediments to the Introduction of the Metrical System of Weights and Measures* (London, 1864); Henry Hennessy, *On a Uniform System of Weights, Measures, and Coins for All Nations* (London, 1862); International Association for Obtaining a Uniform Decimal System of Measures, Weights, and Coins, London, *Concise Narrative of the Origin of the Association* (London, 1856), *Debate in the House of Commons on the Proposed Introduction of the Metric System...1st July, 1863* (London, 1863), *Debates in Both Houses of Parliament on the Metric Weights and Measures Bill: Session 1864* (London, 1864), and *First Report of the Council (February 26, 1857)* (London, 1857); L.A. Lanotte, *The Metric System* (New York, 1867); A. Lanthier, *Universal Measures* (Montreal, 1971); Thomas Leach, *Dozens Versus Tens* (London, 1866); Leone Levi, *On the Metric System of Weights and Measures* (London, 1863) and *The Theory and Practice of the Metric System of Weights and Measures* (London, 1871); B.S. Lyman, *Against*

Adopting the Metric System (Philadelphia, 1897); Justin W. McEachren, International Standardization and an International Decimal System of Weights and Measures (New York, 1917); J. Manning, Future of the Metric and Imperial Systems of Weights and Measures (London, 1899); A.J. Martin, The Metric System in Its Relation to the Surveyor's Profession (London, 1918); E. Noel, Natural Weights and Measures: A Challenge to the Metric System (London, 1889); National Industrial Conference Board, New York, A Digest of "The Metric Versus the English System of Weights and Measures" from Research Report No. 42 (New York, 1921); E.A.W. Phillips, British Trade and the Metric System (London, n.d.); John Scott Porter, On the Metrical System of Weights and Measures, and the Desirableness of Its Universal Adoption (London, 1863); R.T. Rohde, A Decimal System for Great Britain (London, 1885); Samuel B. Ruggles, Report on International Statistical Congress at Berlin in Respect to Uniform Weights, Measures, and Coins (Albany, 1864); Coleman Sellers, The Metric System (Philadelphia, 1880); Gerhard Schlosser, The Decimal System Comparative with Apothecaries Weight (Philadelphia, 1878); E.P. Seaver and G.A. Walton, The Metric System of Weights and Measures (Boston, 1878); Alexander Siemens, Bradford and the Metric System (Manchester, 1913); James E. Stevenson, The Metric System (Newcastle Upon Tyne, 1871); Charles A.L. Totten, An Important Question in Metrology (New York, 1884); D. Neville Wood, Metric Measures for Britain (London, 1969); and James Yates, On the French System of Measures, Weights, and Coins, and Its Adaptation to General Use

(London, 1854).

15. During this era from the 1860s to the First World War there was renewed interest in the metric system in the United States as well, although not nearly as intense as such efforts in Great Britain. For example, the National Academy of Sciences in January of 1867 passed four resolutions favoring increased use of the system in the public and private sectors. In 1870 the House considered four bills widening its use in the government sector. The first enabled persons to use metric units in their business transactions at United States public offices; the second provided that entries of goods at customhouses, and lists and returns for assessment of internal revenue, be made in metric terms; the third and fourth were international coinage system proposals. In 1873 Frederick Barnard, the president of Columbia University, organized the American Metrological Society in New York to lobby for metric adoption, and in 1876 he became president of the American Metric Bureau to disseminate information on the metric system. One year later a resolution of the House requested the executive branch agencies of the government to submit reports concerning the desirability of making the use of the metric system obligatory for all government transactions. On April 15, 1878, the Secretary of the Navy issued an order stipulating that use of the system was obligatory in the medical branch of the Navy Department; in 1894 this was extended to the medical branch of the War Department. Additional metric-sponsored bills appeared in 1895, 1896, 1902, 1906, 1907, and 1918. Aside from

several scientific and medical adoptions, all of these efforts resulted in failure due to legislative mistakes and several well-financed anti-metric campaigns.

Chapter 8

1. The legislation referred to included the Weights and Measures Act of 1904 that amended and supplemented the powers and functions of local inspectors and the Board of Trade, while the Weights and Measures Regulations of 1907 promulgated a list of abbreviations for marking the denominations and capacities of weights and measures; among them were the familiar lb., oz., and cwt. In 1908 Parliament legalized the cran and quarter-cran measures in English and Welsh fresh herrings trade. In 1921 and 1923 Parliament passed the Corn Sales Act and the Fees (Increase) Act respectively. The first established a weight of 60 imperial pounds for the bushel of ground corn. The second declared that the Board of Trade could charge fees for comparing and verifying local standards. The Board and the inspectorate received additional responsibilities by two statutes in 1926—the Weights and Measures (Amendment) Act and the Sale of Food (Weights and Measures) Act. Finally, the Weights and Measures Act of 1936 regulated the measuring, sale, and conveyance of sand, ballast, and similar materials in the United Kingdom. It was the last major innovation in metrological law before 1963.
2. The name was changed in 1920 to the World Metric Standardization

Council, and in 1924 to the All-American Standards Council.

3. Traditionally, the five authorized copies of the yard were called the Parliamentary Copies. The first four were constructed after the 1834 fire and their legal status was maintained by the <u>Weights and Measures Act</u> of 1878. The fifth was constructed in 1879, immediately after the inception of the 1878 act. An order-in-council of August, 1886, approved the latter and placed it in the custody of the Standard Weights and Measures Department of the Board of Trade. These copies are marked: (a) Royal Mint—"Copper 16 oz. Tin 2 1/2 Zinc 1 Mr Baily´s Metal No. 2 STANDARD YARD at 61°.94 Faht. Cast in 1845..." and "No 1 PC 1844 1 lb;" (b) Royal Society—"...No. 3 STANDARD YARD at 62°.10 Faht. Cast in 1845..." and "No 2 PC 1844 1 lb;" (c) Royal Greenwich Observatory—"No. 5 STANDARD YARD at 62°.16 Faht. Cast in 1845..." and "No 3 PC 1844 1 lb;" (d) Palace of Westminster—"...No. 4 STANDARD YARD at 61°.98 Faht. Cast in 1845..." and "No 4 PC 1844 1 lb;" (e) Standard Weights and Measures Department—"Copper 16 oz. Tin 2 1/2 Zinc 1. BAILY´S METAL. PARLIAMENTARY COPY (VI) OF THE IMPERIAL STANDARD YARD. 41 & 42 VICTORIA, CHAPTER 49. STANDARD YARD AT 62° FAHT. CAST IN 1878. Troughton & Simms. London. H.J.C." and "P.C.5 1879."

4. On October 14, 1960, the metric world adopted a new international standard of length that replaced the international meter bar that had served as the standard for over 70 years. This action, taken by the 11th General Conference on Weights and Measures in Paris, defined the meter as 1,650,763.73 wavelengths of the orange-red line of krypton

86, and it rendered obsolete the platinum-iridium meter kept at Paris since the signing of the Treaty of the Meter in 1889. Prior to its adoption, wavelength standards had been suggested by J. Babinet in 1827 (no specific element named), J.C. Maxwell in 1859 (yellow line of sodium), A.A. Michelson in 1890 (red line of natural cadmium), and many more during the first half of the twentieth century. The present definition of the meter is based on a natural constant, the wavelength of a specified kind of light, that scientists believe to be immutable and reproducible with great accuracy in a well-equipped laboratory. It is no longer necessary to return the national standards to Paris at periodic intervals to keep international length measurements on a uniform basis. The meter bars have been retained, however, for use in certain types of measurement and for comparisons between national laboratories. After 1960, the inch became equal to 41,929.399 wavelengths of the krypton light.

5. The fractional figure chosen for the pound in its relation to the kilogram is divisible by 7. Since the grain remained as 1/7000 part of a pound, and the ounce and dram were binary fractions of the pound, there was a finite value in metric terms to the grain and all other imperial units of weight.

6. After the passage of the act the liter was defined for United Kingdom purposes as the volume occupied by a mass of 1 kilogram of pure water at its maximum density and under standard atmospheric pressure. In October, 1964, the General Conference of the International Bureau of

Weights and Measures decided that the liter should be abolished as the scientific unit of volume.

7. Important sources on twentieth-century metric developments are A.H. Hughes, "Problems of Unit Changeover for the United Kingdom," in Systems of Units: National and International Aspects, ed. Carl F. Kayan (Washington, D.C., 1959), pp. 67-83; Mary Baker, Measure for Measure: A Viewpoint on Metrication (London, 1971), pp. 5-22; Ministry of Technology, Change to the Metric System in the United Kingdom (London, 1968), pp. 1-10; United States National Bureau of Standards, Washington, D.C., A Metric America: A Decision Whose Time Has Come (Washington, D.C., 1971), pp. 120-139; Gerald J. Black, Canada Goes Metric, with an Introductory History of Measurement (Toronto, 1974), pp. 135-148; Francis William Kellaway, ed., Metrication (Harmondsworth, England, 1968), pp. 11-95; A.L. LeMaraic, The Complete Metric System with the International System of Units (SI) (New York, 1973), pp. 1-42; Comptes Rendus des séances de la Onzième Conférence Générale des Poids et Mesures: Paris, 11-20 octobre 1960 (Paris, 1961), pp. 91-139; Comptes...de la Treizième Conférence...Paris, etc. (Paris, 1969), pp. 109-120; Irvine C. Gardner, "Wavelength Definition of the Meter," in Systems of Units, etc., pp. 53-63, and British Association for the Advancement of Science, Decimal Coinage and the Metric System: Should Britain Change? (London, 1960), pp. 28-101.

8. See Metrication Board, London, Going Metric: The First 5 Years 1965-1969 (London, 1970) and Going Metric: Progress in 1970 (London,

1971).

Appendix 2

1. Originally a large bundle of more or less cylindrical shape, by the late Middle Ages it had come to designate a closely pressed, rectangularly shaped package, wrapped generally in canvas, and tightly corded or hooped with cotton or iron. See Richard Arnold, Chronicle (London, 1502), p. 206; N.S.B. Gras, The Early English Customs Systems (Cambridge, 1918), p. 697; Hubert Hall, "Select Tracts and Table Books Relating to English Weights and Measures (1100-1742)," in Camden Third Series, 41 (1929), p. 25; Arthur Hopton, A Concordancy of Yeares (London, 1616), p. 164; The Rates of the Custome House Bothe Inwarde and Outwarde (London, 1545), pp. 1 ff.; Noah Bridges, Lux Mercatoria (London, 1660), p. 31; Francis Gouldman, A Copious Dictionary (London, 1664), s.v. bale; E.A. Lewis, ed., The Welsh Port Books (1550-1603) (London, 1927), pp. 97 ff.; Extracts from the Records of the Merchant Adventurers (Durham, 1895), p. 243; Dictionarium rusticum, urbanicum et botanicum (London, 1717), s.v. bale; Joseph Palethorpe, A Commercial Dictionary (Derby, 1829), s.v. bale; "Second Report of the Commissioners...to Consider the Subject of Weights and Measures," in Reports from Commissioners, 7 (1820), p. 6; Richard Rawlyns, Practical Arithmetick (London, 1656), p. 70; William Waterston, A Manual of Commerce (Edinburgh, 1840), p. 147; John Webb, ed., A Roll of the Household Expenses of Richard de Swinfield (London, 1854), p. 80; H.S.

Cobb, ed., The Local Port Book of Southampton for 1439-40 (Southampton, 1961), pp. 63 ff.; and Olive Coleman, ed., The Brokage Book of Southampton: 1443-1444 (Southampton, 1960-61), pp. 1 ff.

2. Just as in the case of the hundred, the hundredweight varied according to product. The most common was 112 pounds (50.802 kg); others were 100 pounds (45.359 kg), 104 pounds (47.173 kg), 108 pounds (48.988 kg), 113 pounds (51.256 kg), 120 pounds (54.431 kg), and 121 pounds (54.884 kg). All of the English weights found in this appendix were based on the avoirdupois system. See Extracts from the Records of the Merchant Adventurers, pp. 57, 100; The York Mercers and Merchant Adventurers: 1356-1917 (Durham, 1918), pp. 156, 168; Hall, pp. 10-11, 18, 22-25, 27-28, 48; George Owen of Henllys, The Description of Penbrokshire (London, 1892), pp. 109, 139; A.L. Merson, ed., The Third Book of Remembrance of Southampton: 1514-1602 (Southampton, 1955), p. 72; Gras, p. 696; The Rates of the Custome House Bothe Inwarde and Outwarde, pp. 4 ff.; Edward Edwards, ed., Liber monasterii de Hyda (London, 1866), p. 68; Churchill Babington et.al., eds., Polychronicon Ranulphi Higden (London, 1865), vol. 1, p. 57; L.T. Smith, ed., A Common-Place Book of the Fifteenth Century (London, 1886), p. 168; Hopton, p. 162; "Second Report," p. 19; Richard Bolton, A Iustice of the Peace for Ireland (Dublin, 1638), p. 274; John Tap, The Path-Way to Knowledge (London, 1613), p. 67; Abraham Rees, ed., The Cyclopaedia (London, 1819), s.v. weights; Samuel Ricard, Traité général du Commerce (Amsterdam, 1781), vol. 2, pp. 152, 155; Dionis Gray, The Store-House

of Breuitie in Woorkes of Arithemetike (London, 1577), p. 7; Dictionarium rusticum, s.v. hundredweight; Malachy Postlethwayt, The Universal Dictionary of Trade and Commerce (London, 1755), vol. 2, p. 188; John Powell, The Assize of Bread (London, 1595), p. C2; M.D. Harris, ed., The Coventry Leet Book (London, 1913), p. 396; Edwardi Bernardi, De mensuris et ponderibus antiquis (Oxford, 1688), pp. 137-138; Thomas Tonkin, ed., Carew´s Survey of Cornwall (London, 1811), p. 45; Humfrey Baker, The Well-Spring of Sciences (London, 1646), p. 211; Politica and Other Tracts (British Museum Manuscript Collections, Sloane 904), folios 212-213; J.H. Alexander, Universal Dictionary of Weights and Measures (Baltimore, 1850), p. 43; L.D.A. Jackson, Modern Metrology (London, 1882), p. 413; and C.W. Pasley, Observations on the Expediency and Practicability of...Improving the Measures, Weights and Money Used in This Country (London, 1834), pp. 113-114.

3. Although the piece was used occasionally for agricultural and metallurgical products, it was employed chiefly for cloth goods. In actual usage, however, the word was pre-empted frequently by "cloth" (or chef, cheef, cheff, chiffe, sheet, caput, etc.) or by the name of the particular fabric. Its length (measured by the yard or ell) and breadth (usually measured by the quarter that equaled 1/4 yard) varied with the quality of the fabric, its construction, its monetary value, and its place of origin or manufacture. Hence, even though the standard piece of cloth was 24 yards (ca. 21.95 m) in length and 7 quarters (ca. 1.60 m) in breadth, there were many exceptions. See

Hall, pp. 11-12; "Report from the Committee Appointed to Inquire into the Original Standards of Weights and Measures in This Kingdom," in Report from Committees of the House of Commons, 2 (1737-65), p. 414; H.G. Richardson and G.O. Sayles, eds., Fleta (London, 1955), p. 120; Gras, pp. 280 ff.; John Strachey et al., eds., Rotuli parliamentorum ut et petitiones (London, 1832), vol. 2, p. 231; The Whole Volume of Statutes at Large (London, 1587), p. 83; A Collection in English of the Statutes Now in Force (London, 1615), p. 465; W. Sheppard, Of the Office of the Clerk of the Market (London, 1665), pp. 45 ff.; Bridges, p. 29; The Acts of the Parliaments of Scotland (London, 1814), vol. 7, p. 252; Ledger of Andrew Halyburton, Conservator of the Privileges of the Scotch Nation in the Netherlands: 1492-1503 (Edinburgh, 1867), p. cxiii; W.E. Lingelbach, The Merchant Adventurers of England (Philadelphia, 1902), p. 111; John Chamberlayne, Magna Britannia Notitia (London, 1708), p. 208; Dictionarium rusticum, s.v. cloth-measure; and "Second Report," p. 27.

4. Additional information on this and other Italian quantity measures may be found in E.H. Byrne, Genoese Shipping in the Twelfth and Thirteenth Centuries (Cambridge, Mass., 1930); F.B. Pegolotti, La Pratica della mercatura, ed. Allan Evans (Cambridge, 1936); Francesco De Luca, Metrologia universale (Naples, 1841); Alfredo Strussi, ed., Zibaldone da Canal: Manoscritto mercantile del sec. XIV (Venice, 1967); Florence Edler, Glossary of Medieval Terms of Business: Italian Series 1200-1600 (Cambridge, 1934); Giuseppe La Mantia, ed., Codice

diplomatico dei Re Aragonesi di Sicilia (Palermo, 1917), vol. 1; C.F. Du Cange, Glossarium mediae et infimae Latinitatis (Paris, 1937), 10 vols.; G. Monticolo and E. Besta, eds., I Capitolari delle Arti Veneziane (Rome, 1896-1914), 3 vols.; D.M. Triulzi, Bilancio dei pesi e misure di tutte le piazze mercantili dell´Europa (Venice, 1803); Angelo Martini, Manuale di metrologia (Turin, 1883); Cesare Manaresi, ed., I Placiti del "Regnum Italiae" (Rome, 1955-60), 3 vols.; Luigi Schiaparelli, ed., I Diplomi di Guido e di Lamberto (Rome, 1906), vol. 1; C. Cipolla and G. Buzzi, eds., Codice diplomatico del monastero di S. Colombano di Bobbio (Rome, 1918), 3 vols.; Armando Petrucci, ed., Il Libro di ricordanze dei Corsini (1362-1457) (Rome, 1965); Christian Bec, ed., Il Libro degli affari proprii di casa de Lapo di Giovanni Niccolini de´Sirigatti (Paris, 1969); Francesco Novati, ed., Epistolario di Caluccio Salutati (Rome, 1911), vol. 4, part 2; Catello Salvati, Misure e pesi nella documentazione storica dell´Italia del Mezzogiorno (Naples, 1970); Bartholomeo di Pasi, Tariffa de pesi e mesure (Venice, 1521); Carlo Afan de Rivera, Tavole di riduzione dei pesi e delle misure delle Due Sicilie (Naples, 1840); Tavole di ragguaglio dei pesi e delle misure (Rome, 1877); and Guérin de Thionville, Tavole delle monete, pesi e misure dei principale paesi del globo (Naples, 1848).

5. See especially Richard Klimpert, Lexicon der Münzen, Masse, Gewichte (Berlin, 1896); Wilhelm Jesse, Quellenbuch zur Münz- und Geldgeschichte des Mittelalters (Halle, 1924); J.P. Heuser, Ueber burgerliche Masse

und Gewichte (Elberfeld, 1839); H. Gloeckner, Allgemeine Münz-, Maass- und Gewichtstabelle (Leipzig, 1887); Horace Doursther, Dictionnaire universel des poids et mesures (Anvers, 1840); Christian Noback, Münz-, Maass- und Gewichtsbuch (Leipzig, 1858); C.W. Rördansz, European Commerce (Boston, 1819); H.J. Alberti, Mass und Gewicht (Berlin, 1957); L.C. Bleibtren, Handbuch des Münz-, Maass- und Gewichtskunde (Stuttgart, 1863); Alphonse Lejeune, Monnaies, poids et mesures des principaux pays du monde (Paris, 1894); Alexis Lemale, Monnaies, poids, mesures et usages commerciaux de tous les états du monde (Paris, 1875); August Blind, Mass-, Münz- und Gewichtswesen (Leipzig, 1906); M.R.B. Gerhardt, Allgemeiner Contorist (Berlin, 1791), 2 vols.; and J.E. Kruse, Allgemeiner und besonders Hamburgischer Contorist (Hamburg, 1784).

6. The carrata originally meant the contents of any cartload.
7. Determined originally as the amount of wood encompassed by a length of cord or string. See Edward Hatton, The Merchant's Magazine (London, 1701), p. 222; James Britten, Old Country and Farming Words (London, 1880), p. 139; The Economist Guide to Weights & Measures (London, 1956), p. 7; George Winter, A Compendious System of Husbandry (London, 1797), p. 101; and Richard Rolt, A New Dictionary of Trade and Commerce (London, 1756), s.v. cord.
8. The livre varied from city to city here as in other French departments. See Jesse, p. 9; Louis Gaillardie, Poids anciens des villes de France (Paris, 1898), pp. 7-9, 25, 31; M.L. Douët-D'Arcq, Comptes de l'hôtel

des rois de France aux XIVe et XVe siècles (Paris, 1865), p. 31; Robertus Senalus, De vera mensvrarvm pondervmqve ratione (Paris, 1535), p. 2; MM. Jourdan et.al., eds., Recueil général des anciennes lois Françaises (Paris, 1825), vol. 13, p. 500; François Garrault, Les Recherches des monnoyes, poix, et manière de nombrer (Paris, 1576), p. 74; Petrus Ciaconius, Opuscula (Rome, 1608), p. 40; Du Cange, s.v. marca; Ephraim Chambers, Cyclopaedia (London, 1728), vol. 2, s.v. weight; Dictionnaire universel Français et Latin (Paris, 1752), s.v. poids; Instruction sur la manière de rectifier les tables de comparaison entre les anciennes et les nouvelles mesures (Paris, 1801), p. 8; G. Bigourdan, Le Système métrique des poids et mesures (Paris, 1901), p. 197; J.B. Duvergier, ed., Collection complète des lois (Paris, 1834), vol. 21, p. 333; C.K. Sanders, A Series of Tables in Which the Weights and Measures of France Are Reduced to the English Standard (London, 1825), p. 13; Doursther, pp. 213-234; Noback, p. 566; Lejeune, pp. 94-95; F. Bailly, "Notice sur les anciennes mesures de Bourgogne," in Société d'histoire...de Beaune (1905), p. 289; P. Guilhiermoz, "Note sur les poids du moyen age," in Bibliothèque de l'Ecole des Chartres, 67 (1906), pp. 162, 188; M.A. Grivel, Les anciennes Mesures de France (Remiremont, 1914), pp. 12, 41-42; Henri Moreau, "The Genesis of the Metric System," in Journal of Chemical Education, 2 (1953), p. 3; Alexander, s.v. livre; L.B.A. Barny de Romanet, Traité historique des poids et mesures (Paris, 1863), p. 13; A.E. Berriman, Historical Metrology (London, 1953), p. 6; M. Noel

Chomel, Dictionnaire oeconomique (Paris, 1740), s.v. poids; Oeuvres complète de Condorcet (Paris, 1804), vol. 18, p. 186; M. Diderot, ed., Encyclopédie...des sciences (Geneva, 1778), vol. 26, p. 422; Encyclopédie méthodique: Commerce (Paris, 1784), vol. 3, pp. 395-411; A.E. Kennelly, Vestiges of Pre-Metric Weights and Measures Persisting in Metric-System Europe (New York, 1928), p. 46; Paul Masson, Histoire du commerce français (Paris, 1911), appendix VIII; A. Mauricet, Des anciennes Mesures de capacité et de superficie dans les départements du Morbihan du Finistère et des Côtes-du-Nord (Vannes, 1893), p. 25; Edouard de Simencourt, Tableaux...des poids et mesures (Paris, 1817), pp. 21, 26; and Philippe Wolff, Commerces et marchands de Toulouse (Paris, 1954), pp. XXVIII-XXIX.

9. The fascicule denoted originally the amount of any material that a man could hold or carry with both arms. See Chomel, s.v. mesure and Bruno Kisch, Scales and Weights: A Historical Outline (New Haven, 1965), p. 1.

10. Originally the fesse was a cord used to bind hay. See H.T. Riley, ed., Memorials of London and London Life (London, 1868), p. 167.

11. The large or great gross was employed principally for the wholesale selling of buttons, beads, cap-hooks, playing cards, various cases and combs, chess pieces, points of thread and silk, and tobacco pipes. See Hall, p. 17; Gras, pp. 696 ff.; The Welsh Port Books, p. 64; Dictionarium rusticum, s.v. gross; "Second Report," p. 18; The Rates of the Custome House Reduced into a Much Better Order (London, 1590), p.

1; Lingelbach, p. 113; Ledger of Andrew Halyburton, p. 288; Hatton, p. 11; Charles King, The British Merchant (London, 1721), vol. 1, p. 282; and The Economist Guide, p. 8.

12. Eight or ten of the sheaves were placed in an upright position; the two remaining sheaves then were placed on top of the others, rising to a peak in the center with their heads slopping downwards at both ends so as to carry off rain. These covering sheaves were called "head-sheaves" or "hoods;" hence the name. See John Ray, A Collection of English Words not Generally Used (London, 1674), p. 24; Britten, p. 146; and Rolt, s.v. hattock.

13. So called from the knot tied around a skein of yarn after reeling.

14. The ream was originally a set of 4 sheets of parchment or paper folded so as to form 8 leaves; this was the unit most commonly used for medieval manuscripts. See L.T. Smith, ed., Expeditions to Prussia and the Holy Land Made by Henry Earl of Derby (London, 1894), p. 159; Hall, p. 25; Hopton, p. 164; Michael Dalton, The Countrey Justice (London, 1635), p. 150; Sheppard, p. 18; Chamberlayne, p. 205; "Second Report," p. 30; C.M. Clode, The Early History of the Guild of Merchant Taylors (London, 1888), p. 307; The Rates of the Custome House Bothe Inwarde and Outwarde, p. 30; Albertus Way, ed., Promptorium Parvulorum (London, 1843), p. 418; Frederick Dinsdale, A Glossary of Provincial Words Used in Teesdale in the County of Durham (London, 1849), p. 100; J.T. Brockett, A Glossary of North Country Words in Use (Newcastle Upon Tyne, 1829), p. 239; The Economist Guide, p. 8; Ledger of Andrew

Halyburton, p. 323; and P.L. Simmonds, The Commercial Dictionary of Trade Products (London, 1883), s.v. ream.

15. The roll was used often in place of the piece as a measure for cloth.

16. The tops of the 15 heads were braided together giving the appearance of a rope. See Hall, p. 28; Sheppard, p. 58; Bridges, p. 30; Britten, p. 175; Simmonds, s.v. rope; and P. Kelly, Metrology: or, an Exposition of Weights and Measures (London, 1816), p. 86.

17. Originally a store or supply of goods.

18. So called because the fur skins were packed and shipped between two heavy boards. The beaver, jennet, miniver, and other furs sometimes were sold singly rather than as a timber. See Henry Spelman, Glossarium Archaiologicum (London, 1664), p. 540; Hall, p. 12; Fleta, p. 120; Gras, p. 166; Expeditions to Prussia, p. 92; A Collection in English of the Statutes, p. 465; Hopton, p. 164; Sheppard, p. 57; Chamberlayne, p. 205; Dictionarium rusticum, s.v. timber; David Macpherson, Annals of Commerce (London, 1805), vol. 1, p. 471; The Rates of the Custome House Reduced, pp. 2 ff.; Statuta tractatus varii registrum brevium (British Museum Manuscript Collections, Add. 32085), folios 150v-151; and Simmonds, s.v. timbre.

Appendix 3

1. See especially M. Diderot, ed., Encyclopédie ou dictionnaire raisonné des sciences (Geneva, 1778), vol. 26, p. 431; Encyclopédie méthodique: Commerce (Paris, 1784), vol. 3, p. 406; P. Kelly, The Universal Cambist

(London, 1821), vol. 1, p. 263; Horace Doursther, _Dictionnaire universel des poids et mesures_ (Anvers, 1840), p. 3; Gennaro Capasso, _Tavole populare delle nuove misure pesi e moneto italiano_ (Naples, 1863), p. 47; Angelo Martini, _Manuale di metrologia_ (Turin, 1883), p. 395; and Catello Salvati, _Misure e pesi nella documentazione storica dell´Italia del Mezzogiorno_ (Naples, 1970), p. 29.

2. It probably referred to a certain mass of stone, lead, or iron that was placed on one scale-pan to determine the weight and, correspondingly, the price of any goods placed on the second pan. See M. Félix Bourquelot, _Etudes sur les Foires de Champagne_ (Paris, 1865), p. 94.

3. See E. Thoison, "Recherches sur les anciennes mesures en usage dans le Gâtinais Seine-et-Marnais et sur leur valeur en mesures métriques," in _Bulletin Historique et Scientifique_, (1903), p. 346.

4. A diminutive of cantaro; see H.J. Alberti, _Mass und Gewicht_ (Berlin, 1957), p. 410 and C. and F. Noback, _Münz- Maass- und Gewichtsbuch_ (Leipzig, 1858), p. 657.

5. The cantaro was a hundredweight for bulkrating wholesale shipments of goods carried long distances overland by sea to foreign markets. It varied generally from 100 to 250 libbre, its exact weight depending on the various local standards for the libbra. See Alberti, pp. 403 ff.; L.C. Bleibtren, _Handbuch der Münz- Maass- und Gewichtskunde_ (Stuttgart, 1863), p. 308; Paul Boiteau, _Les Traités de commerce_ (Paris, 1863), p. 519; W.A. Browne, _The Merchants´ Handbook_ (London, 1899), pp. 445 ff.; E. Cavalli, _Tables de comparaison des mesures, poids et monnaies_

anciens et modernes (Marseille, 1869), p. 57; F.W. Clarke, Weights, Measures, and Money (New York, 1876), p. 107; C. Desimoni, "Observations sur les monnaies, les poids et les mesures cités dans les actes du notaire Génois Lamberto di Sambuceto," in Revue de l'Orient Latin, 3 (1895), p. 22; G.N. Letard, The National Table Book of English & Maltese Weights and Measures (Malta, 1899), p. 28; Doursther, pp. 87-88; Encyclopédie méthodique, pp. 142, 149, 401 ff.; A.E. Kennelly, Vestiges of Pre-Metric Weights and Measures (New York, 1928), pp. 127, 132; Bruno Kisch, Scales and Weights (New Haven, 1965), pp. 229 ff.; Alphonse Lejeune, Monnaies, poids et mesures (Paris, 1894), pp. 207, 211-213; Martini, pp. 33 ff.; Noback, pp. 229 ff.; Salvati, pp. 21 ff.; William Tate, The Modern Cambist (London, 1849), pp. 33 ff.; Giuseppe La Mantia, ed., Codice diplomatico dei Re Aragonesi di Sicilia (Palermo, 1917), vol. 1, pp. 546-548; F.B. Pegolotti, La Pratica della mercatura, ed. Allan Evans (Cambridge, 1936), p. 107; Alfredo Strussi, ed., Zibaldone da Canal: manoscritto mercantile del sec. XIV (Venice, 1967), pp. 48-49; Bartholomeo di Pasi, Tariffa de pesi e mesure (Venice, 1521), p. 56; Ephraim Chambers, Cyclopaedia: or, An Universal Dictionary of Arts and Sciences (London, 1728), s.v. weight; J.M. Benaven, Le Caissier italien (Lyon, 1787), p. 23; D.A. Triulzi, Bilancio dei pesi e misure (Venice, 1803), p. 70; T. Altés, Traité comparatif des monnaies, poids et mesures (Marseille, 1832), p. 292; J.H. Alexander, Universal Dictionary of Weights and Measures (Baltimore, 1850), p. 157; M.G. Canale, Nuova Istoria della repubblica di Genova (Florence, 1860), vol. 3, p. 326; E.H. Byrne, Genoese Shipping in the Twelfth and Thirteenth Centuries (Cambridge, Mass., 1930), p. 11; and F.C. Lane, "Tonnages, Medieval and Modern," in Venice and History: The Collected Papers of Frederic C. Lane (Baltimore, 1966), p. 353.

6. Any Florentine unit designated "after 1836" indicates that it was part of a makeshift system that decimalized the old weights and measures.

7. Any Milanese unit designated "after 1803" indicates that it was part of a new system that represented the metricization of the old weights and measures.

8. This was a derivative of carrata, a cart-load.

9. Permissible subdivisions were 1/2, 1/4, 1/8, 1/16, 1/32, and 1/64. See Wilhelm Jesse, Quellenbuch zur Münz- und Geldgeschichte des Mittelalters (Halle, 1924), p. 135; François Garrault, Les Recherches des monnoyes, poix, et manière de nombrer (Paris, 1576), p. 15; Charles du Fresne Du Cange, Glossarium mediae et infimae Latinitatis (Paris, 1937), s.v. caracca, cayratus; François Le Blanc, Traité historique des monnoyes de France (Amsterdam, 1692), preface; Encyclopédie méthodique, p. 400; Métrologies constitutionnelle et primitive, comparées entre elles et avec la métrologie d'ordonnances (Paris, 1801), vol. 1, p. 112; L. Passot, Tables comparées des anciennes et nouvelles mesures généralement usitées en France (Paris, 1840), p. 97; Noback, p. 567; Lejeune, p. 95; P. Guilhiermoz, "Note sur les poids du moyen age," in Bibliothèque de l'Ecole des Chartres, 67 (1906), p. 181; E. Clemenceau,

Le Service des poids et mesures en France à travers les siècles (Saint-Marcellin-Isère, 1909), p. 180; and M. Denis-Papin, Métrologie générale (Paris, 1946), p. 393.

10. "Poids de marc" (mark weight) refers to the weight system instituted by the government of King John the Good around 1350; also called the "Poids de Paris" in succeeding ages.

11. See G. Monticolo and E. Besta, eds., I Capitolari delle Arti Veneziane (Rome, 1896), vol. 1, pp. 318, 320; Pegolotti, p. 138; Joseph Palethorpe, A Commercial Dictionary (Derby, 1829), p. 17; Zibaldone da Canal, p. 17; Pasi, p. 5; Dominicus Massarius, De Ponderibus & mensuris medicinalibus libri tres (Tiguri apud Froschoverum, 1584), p. 7; Romeo Bocchi, Della guista universale Misura, et suo typo (Venice, 1621), p. 39; F.G. Cristiani, Delle Misure d´ogni genere antiche e moderne (Brescia, 1760), p. 101; Diderot, vol. 26, p. 431; Antonio Menizzi, Parte prima dei pesi delle Stato Veneto (Venice, 1791), p. VIII; Triulzi, p. 143; Florence Edler, Glossary of Medieval Terms of Business: Italian Series 1200-1600 (Cambridge, 1934), p. 321; and Tavole di ragguaglio dei pesi e delle misure già in uso nelle varie provincie del regno (Rome, 1877), pp. 291, 323.

12. See Guiseppe Andreini, Riduzione della misura agraria toscana (Pisa, 1810), p. V; Browne, p. 452; Francesco De Luca, Metrologia universale (Naples, 1841), pp. 87, 157, 196; and Salvati, pp. 37-38. Any Neapolitan units marked "before 1840" or "after 1840" indicate that they were either the old pre-1840 weights and measures or those

representing a decimalization of the old system after 1840.

13. Originally a burden, load, last, or sack for wholesale goods transported to, or sold in, local markets. See MM. Jourdan et al., eds., Recueil général des anciennes lois françaises, depuis l'an 420 jusqu' à la révolution de 1789 (Paris, 1825), vol. 11, p. 606; Du Cange, s.v. chargia, demionus, quarteria; John Harris, Lexicon technicum (London, 1716), s.v. weights; Le Manuel républicain: première partie (Paris, 1799), p. 126; Doursther, p. 99; R. Klimpert, Lexicon der Münzen, Masse, Gewichte (Berlin, 1896), p. 49; and Kisch, p. 242.

14. For the variations from the national standard, see Alexander, s.v. charge; Diderot, vol. 21, p. 682; Doursther, p. 99; Encyclopédie méthodique, pp. 141-142, 151; M.R.B. Gerhardt, Allgemeiner Contorist (Berlin, 1791), vol. 1, p. 373; Martini, pp. 339, 384; and Philippe Wolff, Commerce et marchandes de Toulouse (Paris, 1954), p. XXIX.

15. The word "coppia" meant a pair or couple of anything. See especially Pegolotti, pp. 53-54, 190; Muzio Oddi, Dello Squadro (Milan, 1625), p. 43; and Florence Edler, The Silk Trade of Lucca (Chicago, Unpublished Ph.D. Dissertation, 1930), p. 45.

16. See F. Tomassetti, et al., eds., Statuti della provincia Romana (Rome, 1910), p. 185; Kelly, vol. 1, p. 294; C. Afan de Rivera, Tavole di riduzione dei pesi e delle misure delle Due Sicilie (Naples, 1840), p. 282; Guérin de Thionville, Tavole delle monete, pesi e misure (Naples, 1848), p. 63; Boiteau, p. 519; Martini, p. 395; Browne, p. 445; Kisch,

p. 243; and Salvati, p. 29.

17. See Diderot, vol. 26, p. 420; F. Bailly, "Notice sur les anciennes mesures de Bourgogne," in Société d'histoire, d'archéologie et de littérature de l'arrondissement de Beaune, (1905), p. 296; and Georges Yver, Le Commerce et les marchands dans l'Italie méridionale (Paris, 1903), p. 402. The designation "before 1812" indicates a French unit prior to the establishment of the "Système Usuel."

18. See Louis Gaillardie, Poids anciens des villes de France (Paris, 1898), p. 8; Chambers, s.v. weight; Diderot, vol. 26, p. 422; Bailly, (1905), p. 292; and Kisch, p. 228.

19. See Chambers, s.v. weight; Diderot, vol. 26, p. 420; MM. Berthelot and Laurent, eds., La Grande Encyclopédie (Paris, 1902), vol. 26, p. 1191; Bailly, (1905), p. 295; and M. Geoffroy, Dictionnaire des poids et mesures (Baugé, 1907), p. 318.

20. See Chambers, s.v. weight; Diderot, vol. 26, p. 422; Le Manuel républicain, p. 86; Bailly, (1905), p. 294; and Kisch, p. 228.

21. The denaro was generally synonymous in name with the dramma and in size with the scrupolo. For its historical development, see Jesse, p. 152; Pegolotti, pp. 14, 147, 215; Georg Agricola, Medici libri quinque de mensuris & ponderibus (Paris, 1533), index, s.v. denarius; Massarius, p. 5; Jacques Capelle, De ponderibus, nummis et mensuris (Frankfurt, 1606), p. 24; Bocchi, p. 39; J.J. Manget, Pesi e misure farmacologiche (Pisa, 1703), p. 14; D.M. Manni, "Del Piede Aliprando e del Piede della Porta," in A. Calogerà, Raccotta d'opuscoli, 10 (1734), p. 156;

Cristiani, p. 102; Barnaba Oriani, Istruzione su le misure e su i pesi che si usano nella Repubblica Cisalpina (Milan, 1801), p. 36; Triulzi, p. 116; Alexander, p. 25; Boiteau, p. 519; Browne, pp. 445 ff.; Martini, pp. 17 ff.; De Luca, p. 88; Doursther, p. 122; and Kisch, p. 243.

22. In the monetary system of the Ancient Regime it contained 2 oboles equal to 1/12 sou or 1/240 livre. The designation "before 1800" indicates a French non-metric unit of the Ancient Regime. Of special importance as sources for the weight unit are Recueil général des anciennes lois françaises, vol. 13, p. 500; Garrault, p. 15; Loys Hullin, Le Rapport des poix et monnoyes des anciens aux nostres (Orléans, 1585), p. 15; Du Cange, s.v. marca; Le Blanc, preface; Chambers, s.v. weight; M. Garnier, Histoire de la monnaie depuis les temps de la plus haute antiquité (Paris, 1819), p. 305; Auguste Benoit, Anciennes mesures d´Eure-et-Loir (Chartres, 1843), p. 11; Lejeune, p. 95; Thoison, p. 350; Adolphe Landry, Essai économique sur les mutations des monnaies (Paris, 1910), p. 11; M.A. Grivel, Les anciennes Mesures de France, de Lorraine & de Remiremont (Remiremont, 1914), pp. 12, 56; Denis-Papin, p. 393; and Henri Moreau, "The Genesis of the Metric System," in Journal of Chemical Education, 2 (1953), p. 3.

23. The designation "1800-12" indicates a unit in the special French system of weights and measures that combined the system of the Ancient Regime with the metric system; hence, the old units of livre, once, gros, denier, etc. were retained for use in the provinces, while the metric

system was employed by the government.

24. For a discussion of this unit, see "An Account of the Proportions of the English and French Measures and Weights," in Philosophical Transactions, 42 (1742-43), pp. 187-188.

25. See Robertus Senalus, De vera mensvrarvm pondervmqve ratione (Paris, 1535), p. 2; Garrault, p. 3; Du Cange, s.v. marca; Doursther, p. 133; Browne, p. 454; M. Noel Chomel, Dictionnaire oeconomique (Paris, 1740), s.v. poids; Dictionnaire universel Français et Latin (Paris, 1752), s.v. poids; Martini, p. 473; P. Guilhiermoz, "Remarques diverses sur les poids et mesures du moyen age," in Bibliothèque de l´Ecole des Chartres, 80 (1919), pp. 22, 28; Armando Petrucci, ed., Il Libro di ricordanze dei Corsini (Rome, 1965), p. 124; Massarius, p. 5; Bocchi, p. 40; Lewes Roberts, The Merchants Map of Commerce (London, 1677), p. 298; Manget, p. 14; Diderot, vol. 26, p. 431; J.E. Kruse, Allgemeiner und besonders Hamburgischer Contorist (Hamburg, 1784), p. 318; and Menizzi, p. VIII.

26. See especially M.L. Douët-d´Arcq, Comptes de l´argenterie des rois de France au XIVe siècle (Paris, 1851), pp. 26, 69.

27. For gems and precious metals the grain was subdivided in the mints into units of 1/2, 1/4, 1/8, 1/16, 1/32, 1/64, 1/128, and 1/256. See Edvardus Bernardus, De Mensuris et ponderibus antiquis (Oxoniae, 1688), p. 83; Instruction sur la manière de rectifier les tables de comparaison entre les anciennes et les nouvelles mesures (Paris, 1801), p. 8; and G. Bigourdan, Le Système métrique des poids et mesures

(Paris, 1901), p. 104.

28. Also Bologna, Cagliari, Carrara, Casale Monferrato, Cesena, Crema, Domodossola, Guastalla, Milan, Modena (for silk), Novara, Pallanza, Parma, Pistoia, Tortona, and Urbino.

29. Also Bobbio, Brescia, Como, Genoa, Lodi, Malta, Mortara, Pavia, Piacenza, Porto Maurizio, and Voghera.

30. Also Lucca, Massa, Pesaro, and Senigallia.

31. Also Florence, Lucca (for gold), Modena, Perugia, and Rome.

32. Before 1800 the gros was synonymous with the drachme, and before the sixteenth century it was called a "ternal," "treiseau," or "tresel." See especially Oeuvres complète de Condorcet (Paris, 1804), vol. 18, p. 186; Doursther, p. 163; Bailly, (1905), pp. 295-296; and Guilhiermoz, "Remarques diverses sur les poids," pp. 10-13.

33. The designation "1812-40" indicates a unit in the French "Système Usuel" that established as optional a system of measurements based on metrics that utilized such multiples and submultiples as would make the new units approximately equal to those of the old system.

34. "Libbra nuova italiana" (new Italian pound) refers to the libbra and other weights of the metricized Milanese system after 1803.

35. Sources important for the historical development of the libbra are Luigi Schiaparelli, ed., I Diplomi di Guido e di Lamberto (Rome, 1906), vol. 1, p. 7; C. Cipolla and G. Buzzi, eds., Codice diplomatico del monasterio di S. Colombano di Bobbio (Rome, 1918), vol. 2, pp. 4, 181, 360; Liber magistri Salmonis, sacri palatii notarii: 1222-1226 (Rome,

1906), p. 11; Jesse, p. 65; Statuti della provincia Romana, p. 185; Pegolotti, pp. 14, 161; Vincenzo de Bartholomaeis, ed., Cronaca Aquilana rimata di Buccio di Ranallo (Rome, 1907), p. 181; L. Zdekauer and P. Sella, eds., Statuti di Ascoli Piceno (Rome, 1910), vol. 1, p. 386; Zibaldone da Canal, pp. 17, 68; Il Libro di ricordanze dei Corsini, p. 110; Pasi, p. 19; Senalus, p. 2; Massarius, p. 7; Petrus Ciaconius, Opuscula (Rome, 1608), p. 40; Du Cange, s.v. libra; Benaven, p. 23; Triulzi, p. 14; Capasso, p. 31; Alexander Del Mar, Money and Civilization (London, 1867), p. 9; C.P. Rocca, Pesi e misure antiche di Genova (Genoa, 1871), p. 110; and Kennelly, p. 112.

36. See Jesse, p. 9; Gaillardie, pp. 7-9, 25, 29; M.L. Douët-d'Arcq, Comptes de l'hôtel des rois de France (Paris, 1865), p. 31; Senalus, p. 2; Edouard de Simencourt, Tableaux...des poids et mesures (Paris, 1817), pp. 21, 26; Paul Masson, Histoire du commerce français (Paris, 1911), Appendix VIII; A. Mauricet, Des anciennes Mesures de capacité et de superficie (Vannes, 1893), p. 25; Recueil général des anciennes lois françaises, vol. 13, p. 500; Alexander, s.v. livre; L.B.A. Barny de Romanet, Traité historique des poids et mesures (Paris, 1863), p. 13; A.E. Berriman, Historical Metrology (London, 1953), p. 6; Chomel, s.v. poids; Oeuvres complète de Condorcet, vol. 18, p. 186; Diderot, vol. 26, p. 422; Del Mar, p. 198; Garrault, p. 74; Chambers, s.v. weight; Instruction, p. 8; Bigourdan, p. 197; J.B. Duvergier, ed., Collection complète des lois (Paris, 1834), vol. 21, p. 333; C.K. Sanders, A Series of Tables in Which the Weights and Measures of France Are

Reduced to the English Standard (London, 1825), p. 13; Doursther, pp. 213-231; Noback, pp. 566 ff.; Martini, pp. 64 ff.; Lejeune, pp. 94 ff.; Bailly, (1905), p. 289; Guilhiermoz, "Note sur les poids du moyen age," pp. 162, 188; Grivel, p. 12; Kennelly, p. 46; Moreau, p. 3; and Kisch, pp. 228-230, 232, 248, 251.

37. This was the first national standard of France established by the government of Charlemagne around 800.

38. Instituted by the government of King John the Good, it was subdivided in two different ways: for valuable goods such as gold and silver it consisted of 2 marcs, 16 onces, 128 gros, 384 deniers, or 221,184 primes; for cheaper commodities it was 2 demi-livres, 4 quarterons, 8 demi-quarterons, 16 onces, or 32 demi-onces. The apothecaries used this livre also but subdivided it into 128 drachmes or 384 scrupules.

39. After 1732 it was replaced by the livre poids de marc.

40. The "maille d´esterlin" was part of the poids de marc system.

41. The "maille de denier" was a subdivision of the sou when employed as a weight unit. Both mailles were used interchangeably with obole. See especially J.B.L. Romé de l´Isle, Métrologie (Paris, 1789), p. 207 and Sur l´Uniformité et le système général des poids et mesures (Paris, 1793), p. 35.

42. The local variations contained the same subdivisions as those for Paris. For gold transactions the marc consisted of 24 carats of 32 grains each or 768 grains in all, while for silver it was 12 deniers of 24 grains each or 288 grains in all. See Louis Blancard, "L´Origine du

marc," in Annuaire de la Société de Numismatique (Mâcon, 1888), pp. 224-229.

43. Also Tortona, Cuneo, Alba, Mondovì, Saluzzo, Biella, Vercelli, Turin, Ivrea, Pinerolo, Susa, and throughout Sardinia.

44. Also Milan, Ferrara, and Bergamo.

45. See especially G. Bigwood, "Documents relatifs à une association de marchands italiens aux XIIIe et XIVe siècles," in Académie Royale de Belgique, 77 (1909), p. 239.

46. See especially Casimir Simienowicz, The Great Art of Artillery (London, 1729), pp. 34-35, 64 and C.W. Rördansz, European Commerce (Boston, 1819), p. 501.

47. It was used exclusively for assessing the tonnage of goods transported by ship. See Douët-d'Arcq, Comptes de l'argenterie, p. 36; M. Barreme, Le Livre des comptes (Paris, 1755), p. 10; Le Manuel républicain, p. 98; Noback, p. 567; Bourquelot, p. 79; Martini, p. 473; Bailly, (1905), p. 300; and Kisch, p. 232.

48. Both oboles were used interchangeably with maille.

49. The once for gold and silver consisted of 8 gros, 20 estelins, 40 mailles, or 80 felins; that for pearls and diamonds was made up of 144 carats.

50. See Alexander, p. 77; Maltese Weights and Measures (Board of Trade Manuscript Collections, BT 101 725) (London, 1910), p. 10; Browne, p. 461; Clarke, p. 112; Doursther, p. 371; Letard, p. 28; De Luca, p. 88; Martini, pp. 17 ff.; Pasi, p. 85; Salvati, p. 21; and Tavole di

ragguaglio dei pesi e delle misure, p. 160.

51. See Garrault, p. 76 and Frederic Godefroy, Lexique de l'ancien français (Paris, 1901), s.v. pellet.

52. Variations from approximately 1 1/2 to 14 livres were commonplace throughout France. See Mauricet, pp. 1, 29; Du Cange, s.v. perea, petra; Le Manuel républicain, p. 127; Doursther, p. 396; Bourquelot, p. 94; and Godefroy, s.v. perrée.

53. See Du Cange, s.v. cartaronum, cartayronum, quartaronum, quartonus; Diderot, vol. 26, p. 422; F. Gattey, Eléments du nouveau système métrique (Paris, 1801), p. 99; S. Durant, Tables de comparaison entre les anciens poids et mesures de toutes les communes du département du Gard et les poids et mesures métriques (Nismes, 1816), p. 153; Denis-Papin, p. 393; Wolff, p. XXVIII; and Kisch, p. 232.

54. The designation "after 1840" indicates an official French metric unit after which date other units or systems were illegal.

55. See especially Germain Martin, La Grande Industrie sous le règne de Louis XIV (Paris, 1899), p. 405; August Blind, Mass-, Münz- und Gewichtswesen (Leipzig, 1906), pp. 39-40; and M.D. Dumesnils, Etude critique du système métrique (Paris, 1962), p. 371.

56. See especially R. Fulin, et al., eds., I Diarii di Marino Sanuto (Venice, 1879-1916), vol. 2, p. 154.

57. Also provinces of Avellino, Bari, Benevento, Campobasso, Caserta, Catanzaro, Chieti, Cosenza, Foggia, Lecce, Napoli, Potenza, Reggio di Calabria, Salerno, and Teramo.

58. See especially Christopher Dubost, The Elements of Commerce (London, 1805), vol. 1, pp. 173, 262, 290 and Rapporto della Commissione di Commercio al Gran Consiglio (Milan, 1798), pp. 74-75.

59. Also Ivrea and Pinerolo.

60. Also Acqui, Casale Monferrato, and Capriata d'Orba.

61. The pre-1803 rubbio of 8.170 kilograms also was used at Domodossola, Pallanza, Valsesia, and Vercelli.

62. Also Genoa, Chiavari, Albenga, Savona, La Spezia, Bobbio, Porto Maurizio, and S. Remo.

63. Also Pavia, Mortara, and Sale.

64. The principal unit employed for bulkrating liquids and dry products in wholesale markets or in shipments to regional and interregional ports and depots.

65. By capacity this tonneau was 42 cu pieds (1.440 cu m).

66. The designation "after 1795" indicates a metric unit following the earliest promulgation of the metric system in France by Article 5 of the Law of 18 Germinal, Year III. For the national standards and regional variations, see Alexander, s.v. tonneau; Bailly, (1904), p. 252; Chomel, s.v. mesure; Doursther, pp. 539-544; Encyclopédie méthodique, pp. 124 ff.; Gerhardt, vol. 1, p. 62; Grivel, p. 27; Charles Leroy, Mesures de capacité en usage en Haute-Normandie aux XVIIe et XVIIIe siècles (Rouen, 1937), pp. 8-12; Klimpert, p. 341; Martini, pp. 470-473; Noback, pp. 115, 567; A.J.P. Paucton, Métrologie, ou traité des mesures (Paris, 1780), pp. 799, 811; Malachy

Postlethwayt, The Universal Dictionary of Trade and Commerce (London, 1755), pp. 191-197; Rördansz, pp. 391, 400; and H. Pigeonneau, Histoire du commerce de la France (Paris, 1897), vol. 2, pp. 468, 470.

67. Also the provinces of Piazza Armerina, Gela, Catania, Acireale, Caltagirone, Nicosia, Bivona, Sciacca, Castroreale, Mistretta, Patti, Palermo, Cefalù, Corleone, Termini Imerese, Syracuse, Modica, Noto, Trapani, Alcamo, and Mazara del Vallo.

68. See L'Agence Temporaire des Poids et Mesures (Paris, 1795), p. 11; Le Manuel républicain, pp. 126-127; and Wolff, p. XXVIII.

69. See Le Manuel républicain, p. 127.

INDEX

Index

Index

Index

Index